Predictive Analytics, Data Mining and Big Data

Predictive Analytics, Data Mining and Big Data

Myths, Misconceptions and Methods

Steven Finlay

palgrave
macmillan

First published 2014 by
PALGRAVE MACMILLAN

Palgrave Macmillan in the UK is an imprint of Macmillan Publishers Limited, registered in England, company number 785998, of Houndmills, Basingstoke, Hampshire RG21 6XS.

Palgrave Macmillan in the US is a division of St Martin's Press LLC, 175 Fifth Avenue, New York, NY 10010.

Palgrave Macmillan is the global academic imprint of the above companies and has companies and representatives throughout the world.

Palgrave® and Macmillan® are registered trademarks in the United States, the United Kingdom, Europe and other countries.

ISBN 978–1–137–37927–6

This book is printed on paper suitable for recycling and made from fully managed and sustained forest sources. Logging, pulping and manufacturing processes are expected to conform to the environmental regulations of the country of origin.

A catalogue record for this book is available from the British Library.

Library of Congress Cataloging-in-Publication Data
Finlay, Steven, 1969–
Predictive analytics, data mining and big data : myths, misconceptions and methods / by Steven Finlay.

pages cm

Summary: "Predictive analytics, big data, and data mining are key topics for organizations who want to leverage the ever increasing amounts of data that organizations hold about their customers and other individuals. This in-depth guide provides readers with a solid understanding of data and data trends, the opportunities that it can offer to businesses, the pitfalls and dangers, and a contextual road map for developing solutions that deliver benefits to their organizations. Written in an accessible way, this 'how-to-guide' will help managers to make the most of these technologies in their business area"—Provided by publisher.

ISBN 978–1–137–37927–6 (hardback)

1. Consumer profiling.   2. Consumer behavior—Forecasting.
3. Management—Data processing.   4. Business planning—Data processing.
5. Decision making—Data processing.   6. Data mining.   7. Big data.   I. Title.
HF5415.32.F56 2014
658.8'3402856312—dc23   2014019747

Typeset by MPS Limited, Chennai, India.

To Ruby and Samantha

Contents

Figures and Tables

+

Acknowledgments

First and foremost I would like to thank my wife Samantha and my parents Paul and Ann for their support, comments and proofreading services. I would also like to thank my friend Tracy Moore for providing many useful comments and suggestions on early drafts of the manuscript. Thanks also to the staff of the Management Science Department at Lancaster University in the UK for providing access to the university facilities, which proved invaluable to my writing and research. I am also grateful to the members of the UK Government Operational Research Service (GORS), and in particular my former colleagues in Manchester and Liverpool, for many hours spent chewing over the finer points of predictive analytics, Big Data and life in general during the writing of the book.

chapter 1

Introduction

Retailers, banks, governments, social networking sites, credit reference agencies and telecoms companies, amongst others, hold vast amounts of information about us. They know where we live, what we spend our money on, who our friends and family are, our likes and dislikes, our lifestyles and our opinions. Every year the amount of electronic information about us grows as we increasingly use internet services, social media and smart devices to move more and more of our lives into the online environment.

Until the early 2000s the primary source of individual (consumer) data was the electronic footprints we left behind as we moved through life, such as credit card transactions, online purchases and requests for insurance quotations. This information is required to generate bills, keep accounts up to date, and to provide an audit of the transactions that have occurred between service providers and their customers. In recent years organizations have become increasingly interested in the spaces between our transactions and the paths that led us to the decisions that we made. As we do more things electronically, information that gives insights about our thought processes and the influences that led us to engage in one activity rather than another has become available. A retailer can gain an understanding of why we purchased their product rather than a rival's by examining what route we took before we bought it – what websites did we visit? What other products did we consider? Which reviews did we consult? Similarly, social media provides all sorts of information about ourselves (what we think, who we talk to and what we talk about), and our phones and other devices provide information about where we are and where we've been.

All this information about people is incredibly useful for all sorts of different reasons, but one application in particular is to predict future behavior. By using information about people's lifestyles, movements and past behaviors, organizations can predict what they are likely to do, when they will do it and where that activity will occur. They then use these predictions to tailor how they interact with people. Their reason for doing this is to influence people's behavior, in order to maximize the value of the relationships that they have with them.

In this book I explain how predictive analytics is used to forecast what people are likely to do and how those forecasts are used to decide how to treat people. If your organization uses predictive analytics; if you are wondering whether predictive analytics could improve what you do; or if you want to find out more about how predictive models are constructed and used in practical real-world environments, then this is the book for you.

1.1   What are data mining and predictive analytics?

By the 1980s many organizations found themselves with customer databases that had grown to the point where the amount of data they held had become too large for humans to be able to analyze it on their own. The term "data mining" was coined to describe a range of automated techniques that could be applied to interrogate these databases and make inferences about what the data meant. If you want a concise definition of data mining, then "The analysis of large and complex data sets" is a good place to start.

Many of the tools used to perform data mining are standard statistical methods that have been around for decades, such as linear regression and clustering. However, data mining also includes a wide range of other techniques for analyzing data that grew out of research into artificial intelligence (machine learning), evolutionary computing and game theory.

Data mining is a very broad topic, used for all sorts of things. Detecting patterns in satellite data, anticipating stock price movements, face recognition and forecasting traffic congestion are just a few examples of where data mining is routinely applied. However, the most prolific use of data mining is to identify relationships in data that give an insight into individual preferences, and most importantly, what someone is likely to do in a given scenario.

This is important because if an organization knows what someone is likely to do, then it can tailor its response in order to maximize its own objectives. For commercial organizations the objective is usually to maximize profit.

However, government and other non-profit organizations also have reasons for wanting to know how people are going to behave and then taking action to change or prevent it. For example, tax authorities want to predict who is unlikely to file their tax return correctly, and hence target those individuals for action by tax inspectors. Likewise, political parties want to identify floating voters and then nudge them, using individually tailored communications, to vote for them. Sometime in the mid-2000s the term "predictive analytics" became synonymous with the use of data mining to develop tools to predict the behavior of individuals (or other entities, such as limited companies). Predictive analytics is therefore just a term used to describe the application of data mining to this type of problem.

Predictive analytics is not new. One of the earliest applications was credit scoring,[1] which was first used by the mail order industry in the 1950s to decide who to give credit to. By the mid-1980s credit scoring had become the primary decision-making tool across the financial services industry. When someone applies to borrow money (to take out a loan, a credit card, a mortgage and so on), the lender has to decide whether or not they think that person will repay what they borrow. A lender will only lend to someone if they believe they are creditworthy. At one time all such decisions were made by human underwriters, who reviewed each loan application and made a decision based on their expert opinion. These days, almost all such decisions are made automatically using predictive model(s) that sit within an organization's application processing system.

To construct a credit scoring model, predictive analytics is used to analyze data from thousands of historic loan agreements to identify what characteristics of borrowers were indicative of them being "good" customers who repaid their loans or "bad" customers who defaulted. The relationships that are identified are encapsulated by the model. Having used predictive analytics to construct a model, one can then use the model to make predictions about the future repayment behavior of new loan applicants. If you live in the USA, you have probably come across FICO scores, developed by the FICO Corporation (formerly Fair Isaac Corporation), which are used by many lending institutions to assess applications for credit. Typically, FICO scores range from around 300 to about 850.[2] The higher your score the more creditworthy you are. Similar scores are used by organizations the world over. An example of a credit scoring model (sometimes referred to as a credit scorecard) is shown in Figure 1.1.

To calculate your credit score from the model in Figure 1.1 you start with the constant score of 670. You then go through the scorecard one characteristic at a time, adding or subtracting the points that apply to you,[3] so, if your

Constant	670		
Employment status		**Outstanding mortgage**	
Full-time	28	<$40,000	11
Part-time	7	$40,001–$60,000	0
Homemaker	0	$60,001–$100,000	–9
Retired	15	$100,000–$150,000	–12
Student	–8	$150,000–$250,000	–16
Unemployed	–42	> $250,000	–19
		Not a home owner	0
Time in current employment			
Not in full or part-time employment	0	**Number of credit cards**	
<1 year	–25	0	–17
1–3 years	–10	1–3	12
1–10 years	0	4–8	0
> 10 year	31	9+	–18
Residential status		**Number of days currently past due on**	
Home owner	26	**existing credit agreements**	
Renting	0	0 (All accounts up to date)	14
Living with parent	0	1–30 days past due	0
		31–60 days past due	–29
		>60 days past due	–41
Loan amount requested as			
proportion of annual income			
<10%	43	**Declared bankrupt within the last 5 years?**	
10%–25%	22	Yes	–50
26%–60%	0	No	9
> 60%	–28	Unknown	0

FIGURE 1.1 Loan application model

employment status is full-time you add 28 points to get 698. Then, if your time in current employment is say, two years, you subtract 10 points to get 688. If your residential status is Home Owner you then add 26 points to get 714, and so on.

What does the score mean? For a credit scoring model the higher the score the more likely you are to repay the loan. The lower the score the more likely you are to default, resulting in a loss for the lender. To establish the relationship between score and behavior a sample of several thousand completed loan agreements where the repayment behavior is already known is required. The credit scores for these agreements are then calculated and the results used to generate a score distribution as shown in Figure 1.2.

The score distribution shows the relationship between people's credit score and the odds of them defaulting. At a score of 500 the odds are 1:1. This

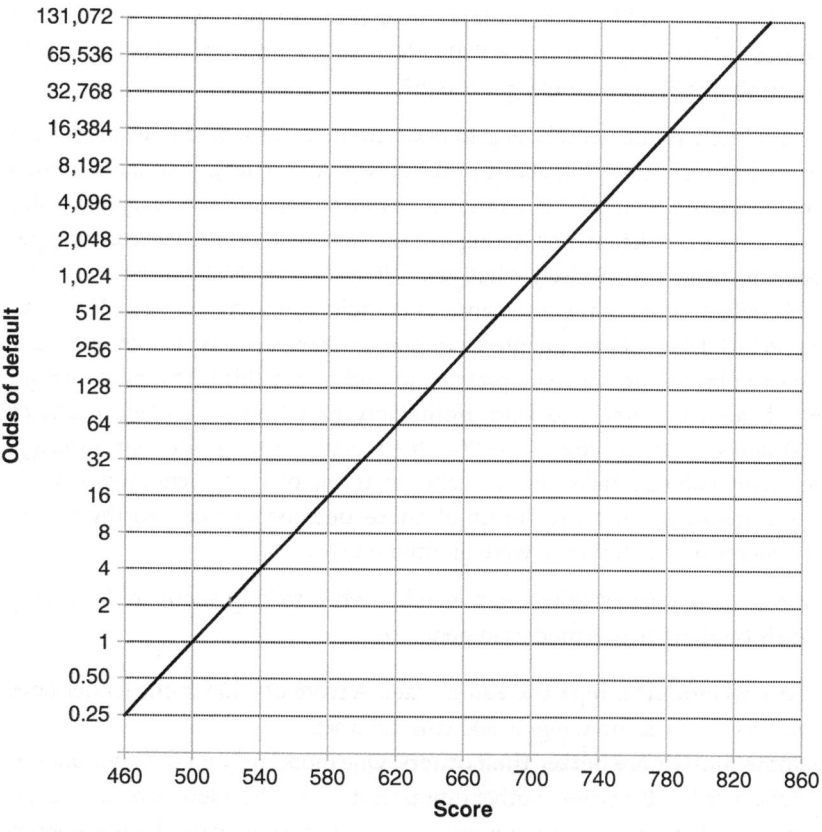

FIGURE 1.2 Score distribution

means that on average half of those who score 500 will default if they are granted a loan. Similarly, for those scoring 620 the odds are 64:1; i.e. if you take 65 borrowers that score 620, the expectation is that 64 will repay what they borrow, but one will not.

To make use of the score distribution in Figure 1.2 you need to have a view about the profitability of loan customers. Let's assume that we have done some analysis of all loan agreements that completed in the last 12 months. This tells us that the average profit from each good loan customer who repaid their loan was $500, but the average loss when someone defaulted was $8,000. From these figures it is possible to work out that we will only make money if there are at least 16 good customers for every one that defaults ($8,000/$500 = 16). This translates into a business decision to offer a customer a loan only if the odds of them being good are more than 16:1. You can see from the score distribution graph that this equates to a *cut-off score* of 580.

Therefore, we should only grant loans to applicants who score more than 580 and decline anything that scores 580 or less. So given the model in Figure 1.1, do you think that you would get a loan?

An absolutely fundamental thing to understand about a predictive model like this is that we are talking about probability, not certainty. Just like a human decision maker, no model of consumer behavior gets it right every time. We are making a prediction, not staring into a crystal ball. Whatever score you get does not determine precisely what you will do. Scoring 800 doesn't mean you won't default, only that your chance of defaulting is very low (1 in 32,768 to be precise). Likewise, for people scoring 560 the expectation is that eight out of every nine will repay – still pretty good odds, but this isn't a pure enough pot of good customers to lend profitability based on an average profit of $500 and an average loss of $8,000. It's worth pointing out that although the credit industry talks about people in terms of being "creditworthy" or "uncreditworthy," in reality most of those deemed uncreditworthy would actually repay a loan if they were granted one.

Some other important things to remember when talking about credit scoring models (and predictive models in general):

- **Not all models adopt the same scale.** A score of 800 for one lender does not mean the same thing as 800 with another.
- **Some models are better than others.** One model may predict your odds of default to be 20:1 while another estimates it to be 50:1. How good a model is at predicting behavior depends on a range of factors, in particular the amount and quality of the data used to construct the model, and the type of model constructed. (Scorecards are a very popular type of model, but there are many other types, such as decision trees, expert systems and neural networks.)
- **Predictions and decisions are not the same thing.** Two lenders may use the same predictive model to calculate the same credit score for someone, but each has a different view of creditworthiness. Odds of 10:1 may be deemed good enough to grant loans by one lender, but another won't advance funds to anyone unless the odds are more than 15:1.

1.2   How good are models at predicting behavior?

In one sense, most predictive models are quite poor at predicting how someone is going to behave. To illustrate this, let's think about a traditional paper-based mail shot. Although in decline, mail shots remain a popular tool employed by marketing professionals to promote products and services to consumers. Consider an insurance company with a marketing strategy that involves sending mail shots to people offering them a really good deal on life insurance. The company uses a response model to predict who is most likely to want life insurance, and these people are mailed.

If the model is a really good one, then the company might be able to identify people with a 1 in 10 chance of taking up the offer – 10 out of every 100 people who are mailed respond. To put it another way, the model will get it right only 10% of the time and get it wrong 90% of the time. That's a pretty high failure rate! However, what you need to consider is what would happen without the model. If you select people from the phone book at random, then a response rate of around 1% is fairly typical for a mail shot of this type. If you look at it this way, then the model is ten times better than a purely random approach – which is not bad at all.

In a lot of ways we are in quite a good place when it comes to predictive models. In many organizations across many industries, predictive models are generating useful predictions and are being used to significantly enhance what those organizations are doing. There is also a rich seam of new applications to which predictive analytics can be applied. However, most models are far from perfect, and there is lots of scope for improvement. In recent years, there have been some improvements in the algorithms that generate predictive models, but these improvements are relatively small compared to the benefits of having more data, better quality data and analyzing this data more effectively. This is the main reason why Big Data is considered such a prize for those organizations that can utilize it.

1.3   What are the benefits of predictive models?

In many walks of life the traditional approach to decision making is for experts in that field to make decisions based on their expert opinion. Continuing with our credit scoring example, there is no reason why local bank managers can't make lending decisions about their customers (which is what they used to do in the days before credit scoring) – one could argue that this would add that personal touch, and an experienced bank manager should be better able to assess the creditworthiness of their customers than some impersonal credit scoring system based at head office. So why use predictive models?

One benefit is speed. When predictive models are used as part of an automated decision-making system, millions of customers can be evaluated and dealt with in just a few seconds. If a bank wants to produce a list of credit card customers who might also be good for a car loan, a predictive model allows this to be undertaken quickly and at almost zero cost. Trawling through all the bank's credit card customers manually to find the good prospects would be completely impractical. Similarly, such systems allow decisions to be made in real time while the customer is on the phone, in branch or online.

A second major benefit of using predictive models is that they generally make better forecasts than their human counterparts. How much better depends on the problem at hand and can be difficult to quantify. However, in my experience, I would expect a well-implemented decision-making system, based on predictive analytics, to make decisions that are about 20–30% more accurate than their human counterparts. In our credit scoring example this translates into granting 20–30% fewer loans to customers who would have defaulted or 20–30% more loans to good customers who will repay, depending upon how one decides to use the model. To put this in terms of raw bottom line benefit, if a bank writes off $500m in bad loans every year, then a reasonable expectation is that this could be reduced by at least $100m, if not more, by using predictive analytics. If we are talking about a marketing department spending $20m on direct marketing to recruit 300,000 new customers each year, then by adopting predictive analytics one would expect to spend about $5m less to recruit the same number of customers. Alternatively, they could expect to recruit about 75,000 more customers for the same $20m spend.

A third benefit is consistency. A given predictive model will always generate the same prediction when presented with the same data. This isn't the case with human decision makers. There is lots of evidence that even the most competent expert will come to very different conclusions and make different decision about something depending on their mood, the time of day, whether they are hungry or not and a host of other factors.[4] Predictive models are simply not influenced by such things. This leads on to questions about the bias that some people display (consciously or unconsciously) against people because of their gender, race, religion age, sexual orientation and so on. This is not to say that predictive models don't display bias towards one group or another, but that where bias exists it is based on clear statistical evidence. Many types of predictive model, such as the scorecard in Figure 1.1, are also explicable. It's easy to understand how someone got the score that they did, and hence why they did or did not get a loan. Working out why a human expert came to a particular decision is not always so easy, especially if it was based on a hunch. Even if the decision maker keeps detailed notes, interpreting what they meant isn't always easy after the event.

Is it important for a predictive model to be explicable? The answer very much depends on what you are using the model for. In some countries, if a customer has their application for credit declined it is a legal requirement to give them an objective reason for the decision. This is one reason why simple models such as those in Figure 1.1 are the norm in credit granting. However, if you are using predictive models in the world of direct marketing, then no one needs to know why they did or didn't get a text offering them a discount on their next purchase. This means that the models can be as simple or as complex as you like (and some can be very complex indeed).

1.4 Applications of predictive analytics

Credit scoring was the first commercial application of predictive analytics (and remains one of the most popular), and by the 1980s the same methods were being applied in other areas of financial services. In their marketing departments, loan and credit card providers started developing models to identify the likelihood of response to a marketing communication, so that only those most likely to be interested in a product were targeted with an offer. This saved huge sums compared to the blanket marketing strategies that went before, and enabled individually tailored communications to be sent to each person based on the score they received. Similarly, in insurance predictive models began to be used to predict the likelihood and value of claims. These predictions were then used to set premiums.

These days, predictive models are used to predict all sorts of things within all sorts of organizations – in fact, almost anywhere where there is a large population of individuals that need decisions to be made about them. The following is just a small selection of some of the other things that predictive models are being used for today:[5]

1. Identifying people who don't pay their taxes.
2. Calculating the probability of having a stroke in the next 10 years.
3. Spotting which credit card transactions are fraudulent.
4. Selecting suspects in criminal cases.
5. Deciding which candidate to offer a job to.
6. Predicting how likely it is that a customer will become bankrupt.
7. Establishing which customers are likely to defect to a rival phone plan when their current contract is up.
8. Producing lists of people who would enjoy going on a date with you.
9. Determining what books, music and films you are likely to purchase next.

10. Predicting how much you are likely to spend at your local supermarket next week.
11. Forecasting life expectancy.
12. Estimating how much someone will spend on their credit card this year.
13. Inferring when someone is likely to be at home (so best time to call them).

The applications of predictive models in the above list fall into two groups. Those in the first group are concerned with yes/no type questions about behavior. Will someone do something or won't they? Will they carryout action A or action B? Models that predict this type of behavior are called classification models. The output of these models (the model score) is a number that represents the probability (the odds)[6] of the behavior occurring. Sometimes the score provides a direct estimate of the likelihood of behavior. For example, a score of 0.4 means the chance of someone having a heart attack in the next five years is 40% (and hence there is a 60% chance of them not having one). In other cases the score is calibrated to a given scale – perhaps 100 means the chance of you having a heart attack is the same as the population average. A score of 200 twice average, a score of 400 four times average and so on. For the scorecard in Figure 1.1, the odds of default double every 20 points – which is a similar scale to the one FICO uses in its credit scores.

All of the first nine examples in the above list can be viewed from a classification perspective (although this may not be obvious at first sight). For example, an online bookseller can build a model by analyzing the text in books that people have bought in the past to predict the books that they subsequently purchased. Once this model exists, then your past purchasing history can be put through the model to generate a score for every book on the bookseller's list. The higher the score, the more likely you are to buy each book. The retailer then markets to you the two or three books that score the most: the ones that you are most likely to be interested in buying.

The second type of predictive model relates to quantities. It's not about whether you are going to do something or not, but the magnitude of what you do. Typically, these equate to "how much" or "how long" type questions. Actuaries use predictive models to predict how long people are going to live, and hence what sort of pension they can expect. Credit card companies build value models to estimate how much revenue each customer is likely to generate. These types of models are called regression models (items 10–13 in the list). Usually, the score from a regression model provides a direct estimate of the quantity of interest. A score of 1,500 generated by a revenue model means that the customer is expected to spend $1,500. However, sometimes what one is interested in is ranking customers, rather than absolute values. The model might be constructed to generate scores in the range 1–100, representing the percentile into which customer spending falls. A score of 1 indicates that the customer is in the lowest spending percentile and a score of 100 that they are in the highest scoring percentile.

In terms of how they look, classification and regression models are very similar, but at a technical level there are subtle differences that determine how models are constructed and used. Classification models are most widely applied, but regression models are increasingly popular because they give a far more granular view of customer behavior. At one time a single credit scoring model would have been used to predict whether or not someone was likely to repay their loan, but these days lenders also create models to predict the expected loss on defaulting loans and the expected revenues from good paying ones. All three models are used in combination to make much more refined lending decisions than could be made by using a single model of loan default on its own.

1.5 Reaping the benefits, avoiding the pitfalls

An organization that implements predictive analytics well can expect to see improvements in its business processes of 20–30% or even more in some cases. However, success is by no means guaranteed. In my first job after graduation, working for a credit reference agency more than 20 years ago, I was involved in building predictive models for a number of clients. In general the projects went pretty well. I delivered good-quality predictive models and our clients were happy with the work I had done and paid accordingly. So I was pretty smug with myself as a hot shot model builder. However, on catching up with my clients months or years later, not everyone had a success story to tell. Many of the models I had developed had been implemented and were delivering real bottom line benefits, but this wasn't universally the case. Some models hadn't been implemented, or the implementation had failed for some reason.

Digging a little deeper it became apparent that it wasn't the models themselves that were at fault. Rather, it was a range of organizational and cultural issues that were the problem. There are lots of reasons why a predictive analytics project can fail, but these can usually be placed into one of three categories:

1. **Not ready for predictive analytics**. Doing something new is risky. People are often unwilling to take the leap of faith required to place trust in automated models rather than human judgment.

2. **The wrong model.** The model builder thought their customer wanted a model to predict one type of consumer behavior, but the customer actually wanted something that predicted a different behavior.
3. **Weak governance.** Implementing a predictive model sometimes requires changes to working practices. As a rule, people don't like change and won't change unless they have to. Just telling them to do something different or issuing a few memos doesn't work. Effective management and enforcement are required.

More than 20 years after I had this realization, methods for constructing predictive models and the mechanisms for implementing predictive models have evolved considerably. Yet I still frequently hear of cases where predictive analytics projects have failed, and it's usually for one of these reasons.

One thing to bear in mind is that different people have different views of what a project entails. For a data scientist working in a technical capacity, a predictive analytics project is about gathering data and then building the best (most predictive) model they can. What happens to the model once they have done their bit is of little concern. Wider issues around implementation, organizational structures and culture are way out of scope.

Sometimes this is fine. If an organization already has an analytics culture and a well-developed analytics infrastructure, then things can be highly automated and hassle-free when it comes to getting models into the business. If the marketing department is simply planning to replace one of its existing response models with a new and a better one, then all that may be involved is hitting the right button in the software to upload the new model into the production environment. However, the vast majority of organizations are not operating their analytics at this level of refinement (although many vendors will tell you that everyone else is, and you need to invest in their technology if you don't want to get left behind). In my experience, it's still typical for model building to account for no more than 10–20% of the time, effort and cost involved in a modeling project. The rest of the effort is involved in doing all the other things that are needed to get the processes in place to be able to use the model operationally.

Even in the financial services industry, where predictive models have been in use longer than anywhere else, there is a huge amount that people have to do around model audit and risk mitigation before a model to predict credit risk can be implemented.[7] What this means in practice is that if you are going to succeed with predictive analytics, you need a good team to deliver the goods. This needs to cover business process, IT, data and organizational culture, with good project management to oversee the lot. Occasionally, a really top class data scientist can take on all of these roles and do everything from the gathering initial requirements through to training staff in how to use the model, but these multi-skilled individuals are rare. More often than not, delivery of analytical solutions is a team effort, requiring input from people from across several different business areas to make it a success.

1.6   What is Big Data?

Large and complex data sets have existed for decades. In one sense Big Data is nothing new, and for some in the industry all the hype around Big Data came as a bit of a surprise. "The emperor's new clothes!" they cried. However, by the early 2010s "Big Data" had become the popular catch-all phrase to describe databases that are not just large, but enormous and complex. There isn't a universally agreed definition of "Big Data," but the features of Big Data[8] that are considered important are:

- **Volume.** Any database that is too large to be comfortably managed on an average PC/laptop/server can be considered Big Data. At the time of writing, Big Data is generally taken to be a database that contains more than a terabyte (1,000 gigabytes) of data.[9] Some Big Data sources contain petabytes of data (1 petabyte = 1,000 terabytes).
- **Variety.** Big Data contains many different types of structured and unstructured data. Structured data is tidy and well defined and can usually be represented as numbers or categories: for example your income, age, gender and marital status. Unstructured data is not well defined. It is often textual and difficult to categorize: for example e-mails, blogs, web pages and transcripts of phone conversations.
- **Volatility (velocity).** Some types of data are relatively static, such as someone's place of birth, gender, social security number and nationality. Other data changes occasionally/slowly, such as one's address, your employer or the number of children that you have. At the other extreme data is changing all the time: for example what music you are listening to right now, the speed you are driving and your heart rate. Big Data is often volatile.
- **Multi-sourced.** Some Big Data sources are generated entirely from an organization's internal systems. This means they have control over its structure and format. However, Big Data often includes external data such as credit reports, census information, GPS data and web pages, and

organizations have little control over how it's supplied and formatted. This introduces additional issues around data quality, privacy and security, over and above what is required from internally sourced data.

This is the sort of definition you usually hear when describing Big Data,[10] but it's difficult to quantify. It doesn't give you a clear dividing line between normal data and Big Data, and therefore many people find it vague and confusing. Using this definition you can't say that a data set is Big Data simply because it contains a particular type of data or a certain amount of data. There is also the onward march of technology to consider. What was Big Data yesterday is not Big Data today, and what's Big Data today may not be Big Data tomorrow – the goal posts are always shifting. There is also a degree of context involved when one talks about Big Data. What constitutes Big Data for a small company, with a few tens of thousands of customers, may be very different from the view held by a large multinational with tens of millions of customers.

Rather than getting hung up on a precise definition of Big Data, an alternative perspective is to view Big Data as a philosophy about how to deal with data, rather than how much data is available or what it contains. The four tenets of this philosophy are:

1. Seek.
2. Store.
3. Analyze.
4. Act.

You proactively search for and obtain new data: you bring all your data together and analyze it to produce insights about what people have done, what they are doing and what they are likely to do in the future. This in turn informs your decision making and what actions to take. A Big Data philosophy is about taking a holistic view of the data available to you and getting the best out of what you have. If an organization is doing this then it doesn't really matter if it has just a few megabytes or many petabytes of data; if the data is structured or unstructured; or where it comes from. From a technology perspective one seeks out IT solutions that deliver the required storage and analytical capability. In some situations this might mean using newer technologies such as Hadoop or Storm, but a traditional relational database solution is often sufficient (or even superior[11]) and should not be ruled out.

So you can view Big Data from a number of perspectives, but for the rest of this book we'll keep things simple and adopt a fairly laid-back definition of

Big Data as: "A very large amount of varied data." So to give a few examples: A government census containing a few dozen items of information about each of that country's 250 million citizens, such as social security number, gender, marital status, income and number of children is a large amount of data, but probably not Big Data. Likewise, the text from 50,000,000 internet pages isn't Big Data either. However, an organization's database of all the information it has gathered about its three million customers in the last five years (purchase details, billing information, e-mails, server logs, texts, transcripts of phone calls, complaint letters, notes taken by staff in branch, credit reports, GPS data and so on) is Big Data.

The first myth I want to dispel is that you need a huge amount of data to build a predictive model. A couple of thousand customer records and a few dozen choice pieces of information about those customers are more than enough, and many useful models have been built using less data than this. However, the more data you have about people, the more predictive your models will be.[12] Big Data has attracted so much interest in recent years because it goes beyond the traditional data sources that people have used in data mining and predictive analytics in the past. In particular, when people talk about Big Data, what they often mean is:

- **Textual data.** This comes from letters, phone transcripts, e-mails, webpages, tweets and so on. This type of data is unstructured and therefore needs a lot of processing power to analyze it.
- **Machine-generated data.** For example, GPS data from people's phones, web logs that track internet usage and telematic devices fitted to cars. Machine-generated data is generally well structured and easy to analyze, but there is a lot of it.
- **Network data.** This is information about people's family, friends and other associates. In some contexts what is important is the structure of the network to which an individual belongs – how many people are in the network, who is at the center of the network and so on. In other contexts the network is a mechanism for inferring things about someone, based on the features of other people in their network.

It used to be the case that the prime source of data for all sorts of predictive models was well-structured internal data sources, possibly augmented by information from a credit reference agency or database marketing company, but these days Big Data that combines traditional data sources with these new types of data is seen as the frontier in terms of consumer information. The problem is that there is so much different and varied data around – so much so that it is becoming increasingly difficult to analyze it all. There is something of an arms race going on between the IT community, who are continuously developing their hardware and software to obtain and store more and more differing and diverse data, and the analytical fraternity, who are trying to find better and more efficient ways to squeeze useful insights from all the data that their IT colleagues have gathered.

1.7   How much value does Big Data add?

When long-established users of predictive analytics, such as banks, insurers and retailers, ask about the value of Big Data, what they really want to know is: What uplift to our predictive models will Big Data provide, over and above the data we use already? I would not go so far as to say that there aren't other applications for Big Data – there are several[13] – but using it to develop predictive models is what people often have in mind. One feature of Big Data is that most of it has a very low information density, making it very difficult to extract useful customer insights from it. A huge proportion of the Big Data out there is absolutely useless when it comes to forecasting consumer behavior. You have to work pretty hard at finding the useful bits that will improve the accuracy of your predictive models – and this is why you need big computers with lots of storage, and clever algorithms, to find the important stuff amongst the chaff.

The way I like to think about this – as have many others – is by an analogy with gold mining.[14] Most industrial activity involved in gold extraction occurs in mines where there are rich seams full of nice big nuggets. Yet it has been estimated that there are thousands of tons of gold dissolved in the world's oceans, possibly more than the entire amount that has been mined across all of human history.[15] However, the gold is very diffuse amongst all that seawater, and no one has yet found a cost-effective way of extracting it. Traditional customer databases are analogous to the gold mine, while Big Data is the ocean. There is lots of useful information floating around in all that data, but it's very sparse compared to traditional data sources and it can be very expensive to get at it.

While we are on the subject of gold, it's worth remembering that in gold rushes (and internet booms and the exciting world of Big Data) the people who sell the tools make a lot of money. Far more strike it rich selling picks and shovels to prospectors than do the prospectors. Likewise, there is a lot of money to be made selling Big Data solutions. Whether the buyer actually gets any benefit from them is not the primary concern of the sales people. There are lots of opportunities for Big Data, but Big Data is not the answer to all of the world's problems and it may not be right for you.

Sometimes the benefits of Big Data can be too small to justify the expense. In banking, for example, the potential for new Big Data sources to improve the predictive ability of credit scoring models is fairly small, over and above the data already available. This is because the key driver of credit risk is past behavior, and the banks have ready access to people's credit reports, plus a wealth of other data, supplied in a nice neat format by Credit Reference Agencies such as Equifax, Experian and TransUnion. At the other end of the spectrum, if you are a marketing team trying to identify people who might be interested in your products and you have nothing to go on, then externally sourced Big Data can provide a lot of value. Being able to trawl the internet, server logs, social network sites, tweets and so on to find out who is talking about what, or what people's friends and family are buying (social network analysis, as discussed in Chapter 9) has immense value over and above the very small amount you would otherwise know about people.

For those already using predictive analytics, one view is that Big Data is very much the icing on the cake once you have good IT systems and good analytics in place – it's a process of evolution not revolution. You can also "bolt on" new technologies such as Hadoop to what you already have. It's not a question of "either/or." However, if your internal data systems are a mess, you don't store as much data as you could and you don't have a strong analytics culture, then Big Data solutions are probably not the next step for you. It's very much a case of trying to run before you can walk. Some organizations have made the leap directly to a Big Data/Analytics culture, but that's rare. It's a high-risk strategy and one that requires commitment, extensive organizational restructuring and a lot of expense. My recommendation is that you should concentrate on getting your own house in order, and making better use of the easy to access data you already have, before moving on to more complex solutions that encompass a wider Big Data philosophy. To put it another way, unless your IT and analytics are already pretty slick, you will get far more bang for your buck from incremental improvements to your current systems, compared to implementing a whole new suite of dedicated hardware and software specifically for handling Big Data.

In terms of the percentage uplift that Big Data provides, that's something of an open question, and is very dependent upon the type of predictive models you want to build and how much data you already make use of. All sorts of figures get bandied about, and as with predictive analytics, you'll only tend to hear about the success stories rather than the failures. Therefore expectations tend to become somewhat over-inflated.[16] Using Big Data to enhance your existing predictive modeling capability is a second-order effect. This means that if you adopt a Big Data approach then you can expect your models to improve, but the improvement won't be anything like as much as when predictive models were implemented for the first time.

My view is that if you already have good data and analytics, and you implement a Big Data Strategy in the right way, then you may see a 4–5% uplift in the performance of your predictive models.[17,18] However, it really depends on the amount of good quality data you already have and what new information your Big Data sources are bringing to the party. If you don't currently have much customer data, and Big Data gives you the ability to predict customer behavior where this wasn't an option before, then you could be looking at benefits of significantly more than 10%. However, if you already have loads of well-organized, accurate and nicely formatted data, then expect to see more modest returns on your investment.

Another perspective, and the one I adhere too, is that the biggest benefits of Big Data have little to do with enhancing existing models in well-run data-rich organizations. Sure, there are some benefits to be had, but the greatest opportunities for Big Data are where it is making new forms of customer prediction viable. One example is where police forces are using huge databases of past crimes to predict the neighborhoods where a crime is likely to be committed in the next few hours. They can then concentrate their resources in those areas.[19] Another area of huge potential is preventative healthcare. Most existing healthcare systems are reactive: they treat you when you are already ill. Combining predictive analytics with Big Data makes it more viable to shift the emphasis to prevention. It becomes possible to predict how likely each citizen is to develop certain conditions and intervene before the illness becomes apparent. This has the potential to add years to average life expectancy.

Marketing is another area where Big Data is proving its worth. For example, by combining information about your movements, gathered from your cell phone, with supermarket data about what type of food you like to buy, you can be targeted with promotional offers for restaurants in the city you are traveling to before you even get there. Another marketing application is to use real-time information about electricity and gas usage to forecast when someone is likely to be at home, and therefore a good time to contact them.

These applications of predictive analytics were little more than science fiction just a few years ago, but this is where the frontier of Big Data and predictive analytics currently lies.

1.8   The rest of the book

I hope this introductory chapter has given you an insight into what Big Data and predictive analytics are, what (one type) of predictive model looks like and how models can be used to help organizations achieve their objectives.

With regard to the rest of the book, Chapters 2–5 are very much about the application and usage of predictive models. Chapter 2 explains how models are used in organizations to drive decision making, and Chapter 3 discusses the analytics culture that needs to be in place if the benefits of predictive analytics are to be realized. In Chapter 4 the focus is on data, in particular the types of data that an organization needs to build good quality predictive models. We also consider the relative value of different types of data, and what data sources an organization should focus on given its current capabilities. There is after all no point getting all excited about Big Data if your existing customer databases are in poor shape. In Chapter 5, we consider ethical and legal issues associated with personal data, and the use of that data within automated decision-making systems built around predictive models.

In the second part of the book, in Chapters 6–10, the focus is more about the process of developing and implementing predictive models. Chapter 6 explains, compares and contrasts the popular types of predictive model that are in use today. This includes linear models, decision trees, expert systems, support vector machines and neural networks. The current trend of generating forecasts of consumer behavior by combining the outputs of several different predictive models (ensemble systems), is also covered.

Chapter 6 is the most technical chapter in the book, but I have attempted to explain everything in a non-technical "formula-free" way without mathematics. However, if it's not your cup of tea, then you can skip this chapter without significant risk of getting lost in subsequent chapters.

Chapters 7 and 8 cover the end-to-end predictive analytics process – what needs to be done to build good quality predictive models and get them implemented within an organization. Chapter 7 discusses each stage of an analytical project, starting with project planning and going right through to implementation within the business and the post-implementation tasks required to ensure that the model continues to generate good quality predictions. Chapter 8 then describes how you go about building a predictive model once you have planned out what you are going to do.

In Chapter 9 we consider two more recent types of data analysis techniques that can be used to enhance the accuracy of predictive models by extracting some of the information that is hidden within Big Data. The first of these is Text Mining (Text Analytics). Text Mining is the art of extracting useful information from speech and text, such as web pages, emails, transcripts of phone conversations and so on. The other method discussed in Chapter 9 is Social Network Analysis. This is about the relationships people have with each other, and how information about our associates can improve the accuracy of predictive models. The final chapter, Chapter10, discusses some of the IT and software issues relating to predictive analytics and Big Data.

2

Using Predictive Models

You may find it surprising, but there are quite a few management books about predictive analytics, data mining and Big Data that *never* explain what a predictive model looks like or how you might actually go about using one. Why, I don't know, but I assume that some people think managers should leave that sort of thing to the experts. Alternatively, some may think that you can't understand these things without getting into the mathematics behind them, but that's not my view.

If you intend to use predictive analytics in your organization, and/or expect to invest heavily in the staff and IT required to deliver high-quality predictive analytics, then it makes sense to have at least some appreciation of what you are investing in and why it will bring benefit. I'm sure you will be glad to hear that this doesn't mean you need to learn any math or statistics. Think about it this way: You don't need to know anything about electronics to operate your TV and you don't need to know anything about mechanical engineering to ride a bike or drive a car. Likewise, it's a myth that you need to have a background in mathematics or statistics to be able to understand how a predictive model works, or how it can be used to improve what you do. If you are comfortable looking at graphs and tables, that's helpful, but that's about as far as it goes.

It's true that using predictive analytics to *build* a predictive model is a technical task, done by nerdy types who enjoy nothing better than discussing equations over a beer (I should know, I used to be one), but understanding what predictive models are and how to use them does not require specialist training. As we saw in the last chapter, a typical credit scoring model, used by banks the world over, works by simply adding up the relevant points to get a score. The higher the score the more creditworthy someone is – it's as simple as that, and don't let anyone try to persuade you otherwise.

2.1 What are your objectives?

If someone starts talking about the predictive models that they could build for you before they have asked you want you want to achieve, then beware. I've lost track of the number of times I have been involved in a project where the conversation starts with a discussion about how we are going to build the model or what data is available to build it with, with little or no thought given to the business problem that the model is going to be used to address.

When discussing predictive models, the starting point should always be some objective within the organization. Predicative analytics then may (or may not) be the right tool to help deliver what you need to do. Models are used to predict all sorts of different things, but whether or not a predictive model is going to help you meet your objectives boils down to just three things:

1. Will the model improve the efficiency of what you do?
2. Will the model result in better decision making?
3. Will the model enable you to do something new that you have not been able to do before?

Let's not beat about the bush. When we talk about efficiency, what we usually mean is replacing a manually based decision-making process with an automated one. Sometimes this results in people being redeployed productively elsewhere, but more often than not efficiency means job losses and/or a devaluation of people's skills. This is important because it means that if you are implementing predictive models for the first time, or are deploying them in a new area where they have not been used before, you will meet resistance and will need a strategy to deal with it. We'll talk more about this issue when we discuss cultural issues around predictive analytics in the next chapter.

With regard to the second point, the evidence from many different studies is that models created using predictive analytics make better predictions than their human counterparts, and in many situations better predictions means making more money. However, having a model that can predict something with a high degree of accuracy is not enough on its own.

Perhaps the biggest mistake people make when developing predictive models is to deliver a model that is not then used for anything. You need to use the predictions generated by the model to do something to influence or control people's behavior, which in turn generates some benefit for them or for you. Identifying people who are likely to purchase something from you is fine, but you then need to act on this information to increase sales. This could be by encouraging your existing customers to spend more (e.g. discount off their next purchase), or to use them as a conduit to attract new customers who would otherwise have spent their money elsewhere (e.g. two for one deal for the customer and a friend).

If we go back to our loan example of Chapter 1, the model predicts the likelihood of default on a loan. Knowing how likely someone is to default does not on its own make you any money. The decision is whether or not to offer loans. Therefore we use the model score as the basis of our decision about whether or not to give someone a loan. Making the right decision on the basis of the score is far more important than the score itself.

When thinking about predictive analytics, always have these three things in mind: How will a predictive model improve operational efficiency? How will it improve your decision making process? And what value does the model provide? If you can't answer these questions then you should reconsider whether it's worth proceeding. Otherwise you risk wasting a lot of time and money creating something that might be very predictive, but doesn't do anything to help you achieve your goals.

2.2   Decision making

In Chapter 1 we introduced a score distribution that a lender used to decide whether or not to grant loans to customers who had applied for one. The score distribution is the key tool that underpins the use of all predictive models and is the basis for assessing how well a model performs. This includes the widely used gain charts and lift charts (see Appendix C for an explanation of what these are). In this chapter we are going to present some further examples of score distributions, and show how an organization would use these to drive their strategies for dealing with consumers. To do this we are going to work through a case study where a predictive model is constructed and used to help an organization meet its objectives.

Booles (not its real name) is an upmarket regional supermarket chain that has decided to set up a national online wine store. It wants to do this because it makes an above-average profit on the wine it sells in-store, and it sees this as a low-cost, low-risk strategy for reaching a much larger customer base than it currently has access to.

The set-up and running costs will be low because Booles can make use of its existing warehousing and IT systems. In fact, the IT department has already set up a prototype website for the wine store. All that remains to be done is to find the right customers to target and persuade them to buy.

A Unique Selling Point (USP) for the wine store is that the company has exclusive access to a large supply of high-quality locally sourced wines that the national retailers don't have, and which are already being sold in Booles' stores. The biggest challenge is to identify which customers outside of the region to target, and then develop an appropriate marketing campaign to attract them to buy online.

So let's start by thinking about what the business wants to achieve with the wine store. The board has laid out the following objectives:

- **Objective 1.** Recruit at least 25,000 customers in the first year of operation.
- **Objective 2.** Make a profit during the first year. Operational overheads have been absorbed by the existing retailing operation. Therefore, the revenues generated from wine sales only need to cover the cost of the wine plus the marketing cost of acquiring new customers.

The board has given the marketing team a budget of $1.2m to meet the objectives. So what sort of data does the marketing team have to play with? Here are the key facts:

- Analysis of in-store till transactions shows that the average profit on a 12 bottle case of wine is $75.
- A direct marketing campaign costs $2 for each person targeted. A typical campaign includes texts, e-mails, mail shots and voice messages, delivered over several weeks.
- The marketing department has access to a contact list, supplied by a database marketing company, containing details of 5.1 million people. The list contains names and contact details, plus geo-demographic information such as income, occupation, age, car ownership and so on. However, the list does not contain information about wine-buying behavior.

At $2 a shot and with $1.2m to spend, the company could run a direct marketing campaign targeted at 600,000 people, selected at random from the contact list. Another option would be to forget the list, and go for a mass communication strategy, spending the recruitment budget on TV ads, sponsorship of websites, articles in magazines and so on. However, the Marketing Director knows that neither of these strategies will be successful. Why? Because quality wines are the domain of a very specific consumer segment. A blanket marketing strategy would not be cost-effective because most of the material would fall on deaf ears. In order to meet the business objective, the Marketing Director knows that they need a way to target just those people who like quality wines, and he believes predictive analytics may be the tool to help them do this. So in this case, predictive analytics is going to be used as an efficiency tool to reduce the costs associated with targeting and recruiting profitable new customers.

So what does the marketing department do? If they want to use predictive analytics then they need some consumer data to analyze. To construct a model using predictive analytics, two types of data are required:

1. **Predictor data (predictor variables)**. This is the information that is going to be used to make the prediction; i.e. the things that could feature in the model. For this problem, the predictor data is the geo-demographic information such as income, occupation, age and so on, that has been supplied with the contact list.
2. **Behavioral (outcome) data.** This is information about the behavior we want to predict. For this example, the behavior is whether the customer buys wine or not. In technical terms, this is what a data scientist calls the "Dependent variable," "Target variable" or the "Modeling objective."

Predictive analytics is all about understanding the relationships between the predictor data and behavioral (outcome) data. You can't do predictive analytics if one of these types of data is missing. For behavioral data, you also need representative samples of each type of behavior so that the differences between behaviors can be analyzed. For our case study, this means that information about people who *did not* buy wine is just as important as information about those that did. The predictive analytics process then analyzes the data to identify how the predictor data can be used to differentiate between each behavior.

The marketing department has lots of predictor data that was supplied with the contact list, but it has no information about outcomes; i.e. wine-buying behavior. Therefore the marketing team needs to obtain some before it can construct a predictive model. To obtain data about wine-buying behavior, the marketing team undertakes a test campaign targeting 100,000 people selected at random from the contact list. At $2 a time, the trial costs $200,000, leaving them with a remaining budget of $1 million.

As the test campaign progresses, the first sales come within just a few hours. Sales peak after about a week, followed by a gradual decline over several weeks. At the end of the sixth week the campaign winds up as new sales drop to zero. Table 2.1 summarizes the key findings from the test campaign.

Table 2.1 provides some very useful information. Out of 100,000 people who were contacted, 1,600 responded by buying wine. This is *a **response rate*** of 1.6%. Each contact cost $2. Therefore the response cost (the average spend required to secure one sale) is $125 (row six in Table 2.1).

The information in Table 2.1 supports the Marketing Director's belief that a random targeting strategy would be unsuccessful. This is because in order to generate the required 25,000 sales specified for objective 1, it would be necessary to target 1,562,500 people (row 7) at a cost of more than $3m (row 8). However, with a $1.2m budget they can only afford to target 600,000. In fact, with a response rate of 1.6%, a random contact strategy can be expected to generate just 9,600 sales from 600,000 contacts (600,000 × 1.6%).

From Table 2.1 we can also establish that a random contact strategy would be loss-making. This is because the average profit of $75 per case won't cover the average $125 it costs to recruit each customer: Overall, they will lose $50 on each case they sell. So the marketing department would fail to meet the second objective as well.

Table 2.1   Key findings for the test campaign

Row	Test campaign: Key statistics	
	Statistic	**Value**
1	Number targeted	100,000
2	Cost of targeting each person	$2
3	Total campaign cost	$200,000
4	Cases of wine sold (number of responses)	1,600
5	Response rate (1,600 ÷ 100,000)	1.6%
6	Response cost ($2 ÷ 1.6%)	$125
7	Number that need to be targeted to sell 25,000 cases of wine (25,000 ÷ 1.6%)	1,562,500
8	Cost of recruiting 25,000 customers (1,562,500 × $2)	$3,125,000

This may seem a little disappointing, but remember that the main purpose of the test campaign is not to generate sales, but to gather data about wine-buying behavior. The marketing department now has some behavioral data to work with and sets about the task of building a predictive model (a response model) using the data it has acquired from the test campaign.

To build the model, the marketing team's data scientist takes all of the available information about the 100,000 people contacted as part of the trial, and loads it into a statistical software package for analysis. There are many specialist software packages that one can use to build predictive models, but the most popular ones are SAS, SPSS and R. These days it's pretty rare to actually do any math yourself when building a predictive model – the software takes care of all the required calculation to generate a model using the appropriate mathematical/statistical technique.

There are several different types of predictive models and literally dozens of different algorithms that can be used to construct a predictive model. Each has its own strengths and weaknesses, and we describe these in Chapter 6. Each method of model construction also has various parameters that can be set, so in theory there are millions of possible models that could be constructed for any given problem.

The role of the data scientist is to decide what technique/parameters to use and explore the range of possible models, using their experience and expertise to derive the best model that they can. What one means by "best" varies from project to project. Obviously you want the model to be as predictive as possible, but often there are business requirements and constraints that need to be taken into account. It is very common to sacrifice a small amount of predictive accuracy to ensure that these business requirements are met. For example, in many industries there is a desire for models to be simple and explicable. This is so that the way the model calculates its forecast can be understood by non-experts and the models are easy to implement – just like the loan scorecard we discussed in Chapter 1.

Sometimes the model must look a certain way and conform to business expectation. Similarly, a model developer may be required to force certain variables to feature in the model and/or ensure that certain ones are excluded.[1] Data scientists often have to play these requirements off against each other, and this is one area where model building becomes something of an art – going beyond pure statistics and overlaying business knowledge to deliver something that is predictive, useable and adds value to what you do.

Returning to our case study, after a week of developing models, tweaking them and exploring different options, the data scientist comes up with what they think is a really good model, as shown in Figure 2.1

This type of model is called a "Decision Tree." You will notice it looks very different from the scorecard type model introduced in Chapter 1 for calculating credit scores. With the decision tree you calculate a score as follows: A case begins at the top of the tree. The case then goes down one of the branches, depending on its attributes. The process is then repeated at the next branching and so on, until the case eventually ends up at one (and only one) of the end nodes[2] of the tree.

In Figure 2.1 the end nodes are shaded in grey and numbered from 1 to 11. If we take someone aged 30, who is male and a homeowner, they fall into end node 5. Likewise, someone aged 50, with one child, married and with an income of less than $40,000, falls into end node 7. If you remember back to the scorecard in Chapter 1, an individual's score was calculated by adding up the points that they received for each of their attributes. With a decision tree the score is simply the end node into which an individual falls.[3] So the decision tree in Figure 2.1 generates scores ranging from 1 to 11 depending upon the node in which someone ends up.

To examine how well the tree predicts wine buying behavior, the marketing department puts each of the 100,000 cases from the test mailing through the

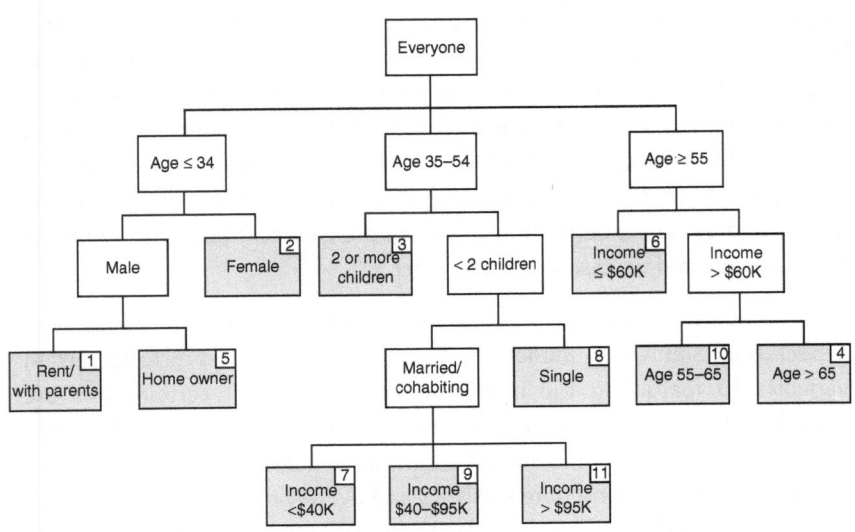

FIGURE 2.1 A decision tree

tree[4] and sees how many cases get each score, and most importantly, the number of people with that score who did or did not buy wine. They then summarize the results in the form of a score distribution table, as shown in Table 2.2.

The score distribution has two parts. We will begin by looking at the columns on the left-hand side. These show what happened to the 100,000 people targeted as part of the test campaign. Column A shows the score. Column B shows the number of people with each score who were contacted as part of the campaign, and Column C the number who subsequently bought wine. Columns D is the response rate, calculated from columns B and C.

Column E shows the cost per response, i.e. the marketing spend per sale ($2/ Column D), and Column F shows the expected profit from mailing someone with that score, i.e. the $75 gross profit from selling a case minus the cost per response (Column E).

The first thing to note is that the decision tree provides a good degree of discrimination between those who bought wine and those who did not. For customers with a score of one, only one in 200 (0.5%) bought wine – these are clearly not the type of people to target. At the other end of the score distribution, those scoring 11 have a response rate of 8.0%. To put it another way, the odds of people with a score of 11 buying wine are more than 1 in 12. These people are five times more likely to buy wine than people contacted at random and are therefore, very much the sort of people who should be targeted. Using the score distribution (Column F) it's easy to see that targeting those scoring seven or more will be profitable, but targeting anyone scoring six or less will result in a loss.

The second part of the table shows us what the marketing team think will happen when they apply the decision tree to the remaining five million people on the contact list who have not yet been targeted. The data scientist begins by putting the remaining five million cases from the contact list through the decision tree to generate a score for each of them. This is then used to calculate how many people get each score (Column G). Column H gives the total cost of targeting everyone with that score, based on a targeting cost of $2 per person.

The data scientist then infers the expected number of responses from what was observed from the test campaign (Column I). Take those scoring 11 as an example. If the 8.0% response rate observed for the test campaign is applied to the remaining 80,040 people scoring 11, then an estimated 6,403 sales would result (8% × 80,040) and it would cost $160,080 to target these people at $2 a time. The expected profit from mailing everyone with a score of 11 is then calculated to be $320,150 (Column J).[5]

Table 2.2 Score distribution

A	B	C	D	E	F	Remaining population			
	Test campaign								
Score (end node)	Number contacted	Number responded (bought wine)	Response rate (100% × C/B)	Cost per response ($2 / D)	Profit per response ($75 – E)	Number contacted	Cost of targeting (G × $2)	Expected responders (G × D)	Expected profit (I × F)
1	15,617	78	0.500%	$400	–$325	780,850	$1,561,700	3,904	–$1,268,881
2	16,820	103	0.613%	$326	–$251	840,995	$1,681,990	5,159	–$1,295,041
3	13,702	113	0.823%	$243	–$168	685,088	$1,370,176	5,639	–$947,282
4	13,800	167	1.212%	$165	–$90	690,006	$1,380,012	8,364	–$752,734
5	11,139	186	1.667%	$120	–$45	556,930	$1,113,860	9,282	–417,698
6	8,500	191	2.247%	$89	–$14	424,977	$849,954	9,550	–$133,701
7	6,452	177	2.740%	$73	$2	322,605	$645,210	8,838	$17,677
8	3,982	117	2.941%	$68	$7	199,100	$398,200	5,856	$40,991
9	4,736	166	3.509%	$57	$18	236,807	$473,614	8,309	$149,562
10	3,652	174	4.762%	$42	$33	182,602	$365,204	8,695	$286,946
11	1,601	128	8.000%	$25	$50	80,040	$160,080	6,403	$320,150
Total	100,000	1,600	1.600%	$125	–$50.0	5,000,000	10,000,000	80,000	–$4,000,000

To decide how to spend their remaining $1m budget, the marketing team uses Columns G and H to calculate that they can afford to mail all 499,449 people with a score of nine or more at a cost of $998,898, which will generate an estimated 23,407 sales. This is in addition to the 1,600 sales from the test campaign making 25,007 in total.[6] So we say that the cut-off score that the marketing team are going to use for this model is nine. Those scoring nine or above are targeted, those scoring below the cut-off (eight or less) are not.

With 25,007 cases generating an average gross profit of $75 per case and $1,198,898 spent on marketing, the overall net profit[7] is estimated to be $676,627. So it looks like the decision tree will have helped the marketing team meet both of the objectives set by the board. The marketing director decides to spend the remaining $1,102 of the marketing budget on a night out for the team to celebrate.

2.3   The next challenge

Booles runs its campaign, and within six months has met its recruitment target for new customers to its online wine store. The marketing director breathes a huge sigh of relief and gets a decent bonus for his efforts. The board is also pleased; so pleased in fact that they set the marketing team a new target; to generate another $500,000 profit contribution by the end of the year. The board thinks this shouldn't be too much of a challenge given that the first cohort of customers generated a lot more than this. They don't care how much is spent on marketing or how many customers are recruited, as long as the profit target is met.

The problem for the marketing team is that they don't have a new contact list available. They are going to have to make do with the contact list that they already have. However, the best prospects, those that scored nine or more on the decision tree, have all been targeted and there is little value targeting them again.

The first thing the marketing director does is to go back to the score distribution (Table 2.2) to see if it's possible to do something with the decision tree. However, the profitability of the people that have not yet been targeted is low because the response costs are high. Only those scoring seven or eight make any profit at all (Column F), and if the marketing team targeted everyone scoring seven or eight then the expected profit is only $58,668 ($17,677 + $40,991) – well short of the new target.

The team knows that the average profit on a box of wine is $75. However, when they look at the figures in more detail they see that there is quite a lot of variation. The average profit is $75, but some customers are buying expensive wines and generating more than three times that amount. Likewise, those selecting the cheapest wines are generating almost no profit at all. What if they could identify those that make the big profits? If they could do this, then it might be profitable to target some of those with low scores, because the higher profit per case would offset the increased response cost. On this basis the marketing team decides to build a second predictive model, this time to estimate the gross profit on each case of wine sold.

The decision tree model constructed to predict response to marketing activity was a classification model – do people respond by buying wine or not? This time the team is going to construct a regression model to predict the gross profit that Booles makes on each case of wine it sells. Given they have already sold more than 25,000 cases of wine they don't need to do any more test campaigns. They can construct a model using the data they already have to predict the gross profit on a case of wine, *given* that someone responded.

For expediency I won't go into the specific details of the structure of the gross profit model, but the model allocates scores ranging from one to three, as shown in Table 2.3:[8]

The first column in Table 2.3 shows the score. The second column shows the gross profit given the score, and the final column the expected number of responders. Now that the marketing team has two models, it can use them in combination to segment the population, as shown in Table 2.4.

Table 2.4 shows how the initial response score generated from the decision tree are broken out by the scores from the gross profit model. What Table 2.4 tells us is that it is actually profitable to target some people with response scores of

Table 2.3   Score distribution for the gross profit model

Score	Gross profit	Expected number of responders*
1	$29	18,862
2	$74	30,712
3	$203	7,018
Total	**$75**	**56,592**

* The total expected number of responders (56,592) is the sum of the expected responders in Column I of Table 2.2, scoring eight or below.

four, five and six, even though the average profit of people with these scores is negative. Likewise, some scoring seven and eight are loss-making, and should therefore not be targeted as part of the contact strategy.

The cells highlighted in grey are those where the net profit is positive. The white ones are where the net profit is negative. If you sum up the net profit from the shaded cells you will see that the total expected profit from targeting these groups is $518,011. So by using this more refined dual model strategy, the marketing team will once again achieve its objectives.

Table 2.4   Using two models in combination

Score	Response cost	Gross profit	Gross profit score			
			1	2	3	Total
			$29	$74	$203	$75
1	$400	Number	1,209	2,244	452	3,904
		Profit	-$448,539	-$731,309	-$89,044	-$1,268,881
2	$326	Number	2,594	1,621	945	5,159
		Profit	-$770,548	-$408,248	-$116,244	-$1,295,041
3	$243	Number	1,544	3,514	582	5,639
		Profit	-$330,218	-$593,783	-$23,280	-$947,282
4	$165	Number	2,741	4,603	1,021	8,364
		Profit	-$372,780	-$418,751	$38,798	-$752,734
5	$120	Number	2,960	5,220	1,104	9,282
		Profit	-$269,225	-$240,104	$91,632	-$417,698
6	$89	Number	3,435	4,844	1,272	9,550
		Profit	-$206,051	-$72,657	$145,008	-$133,701
7	$73	Number	2,614	5,246	980	8,838
		Profit	-$114,968	$5,246	$127,400	$17,677
8	$68	Number	1,769	3,426	662	5,856
		Profit	-$68,936	$20,558	$89,370	$40,991
Total		Number	18,862	30,712	7,018	56,592
		Profit	-$2,581,253	-$2,439,053	$263,640	-$4,756,669

1. Each cell in the table shows the expected number of responders in the cell and the total profit from the responders in the cell. Total net profit is calculated as:

$$\text{Profit} = (\text{Gross profit} - \text{response cost}) \times \text{number of responders}$$

For example, for a response score of 1 and a gross profit score of 1 the total net profit is calculated as:

$$\text{Total net profit} = (\$29 - \$400) \times 1{,}209 = -\$448{,}539$$

2. The total responders and total profit figures are taken from Columns I and J in Table 2.2.

2.4   Discussion

If you managed to get your head round the first part of the case study then you have gained a good understanding of how predictive models are used. If you were also comfortable with the second part then that's as advanced as many organizations get.

In some situations there may be more models, or more complex models predicting all sorts of different outcomes, but it's all the same sort of thing with some minor variations. Models generate scores that predict behavior. Cut-off score(s) are then set by the business to determine how people are treated given their score. People scoring above the cut-off are treated one way, those below the cut-off are treated in another. This applies whether we are talking about a bookseller using a predictive model to identify the books you are most likely to buy, a doctor using a model to decide whether or not to send you for further tests, or the government using predictive analytics to hunt down tax evaders. It's all about using the scores generated by the model to identify the people of interest, and then making the right decisions about what to do with those people once you have identified them.

In the Booles example the marketing department combined its two models. It did this by cross-tabulating the two scores and then making decisions about people depending upon the cell in which they fell. This is a common, very simple and transparent method for using several model scores together to make a decision, when each model predicts a different type of behavior. However, it's not the only way to combine scores.

Another popular method of using two (or more) models in tandem is to create a new score by multiplying the individual scores together. Decisions are then made on the basis of the new score. This works particularly well when one model represents a probability of a behavior (such as response likelihood) and the other model captures the magnitude of that behavior (such as the gross profit on the wine). By multiplying the two scores together you get an estimate of the expected magnitude of the behavior prior to you targeting them.[9] If the likelihood of response is 10% and the predicted gross profit on each wine sale is $50 then the expected net profit, prior to targeting, is $5 (10% × $50). If your objective is profit maximization, then all you do is simply target everyone where the expected gross profit is greater than cost of targeting them, i.e. the expected net profit is greater than zero.

Another example of dual model usage is setting insurance premiums. An insurer will build a classification model to predict how likely someone is to make a claim and a regression model to estimate the value of the claim, should a claim be made. To set the value of someone's premium the insurer multiplies the likelihood of a claim by the estimated value of the claim. They then add an amount to cover their administrative costs and profit margin to arrive at a premium.

If you want to get really clever when it comes to marketing insurance, you can extend this idea further and construct models to predict many different things along the customer engagement path. You could, for example, build models to predict all of the following:

- The likelihood of someone responding to marketing activity by requesting an insurance quote.
- The likelihood of someone who already has a quote taking up the insurance offered.
- The likelihood of the insured making a claim.
- The value of the claim, should a claim be made.
- The expected legal and administrative costs of a claim, should a claim be made.

You could then use all five models in combination to generate quotes for customers. If you want to go one step further, then you also build a price sensitivity model and adjust the price of insurance to the point just below where your customer decides to shop elsewhere.

Coming back to Booles, a good question to ask is: Why did the marketing team begin by building a model to predict response, rather than a model to predict the profitability of customers? The reason is that that this was not what they were asked to do. Their original objective was to obtain a certain number of new customers. Consequently, they built a model to identify any potential customers, not the most profitable ones. Profitably was a constraint (must be greater than zero), but maximizing profit was not an objective for them.

OK, but why didn't they build a single profit model for the second part of the problem, when they were asked to generate profit? Why did they use two models to predict response and gross profit respectively? For problems like this, you usually get better predictions from building several component models and then combining them, instead of building a single model to predict the final outcome. This is particularly true when there is a "hurdle" to overcome. For Booles, the hurdle is response. 98.4% of people targeted failed the hurdle, i.e. generated zero profit because they did not respond. If all these cases where included in the sample used to build the gross profit model, then the modeling process would be dominated by zero cases. The practical result would be that the model would be very good at predicting zero cases, but not anything like as good at picking out the relatively few high-profit ones.

For hurdle problems like this, the larger the proportion that fail the hurdle, the better it is to split up the problem and have two models. The first model predicts whether the individual will pass the hurdle, while the second model predicts the magnitude of the behavior exhibited by those that pass the hurdle.

2.5   Override rules (business rules)

The Booles case study demonstrates a very purist use of predictive models. Decisions were made using only the scores from the models – no human involvement was required in determining what the outcomes should be. One perspective is that this is how you should use your models to get the best from them. Each and every decision about a customer is made automatically on the basis of the score alone. However, in many practical applications it is common to use business rules to *override* score-based decisions in certain circumstances. These override rules are applied to meet strategic objectives beyond the scope of the model, or to ensure that certain actions are/are not taken for some individuals, regardless of the score that they receive.

One reason for override rules is legislation. There are laws that require you to treat certain people in certain ways; otherwise you face the risk of legal action and/or reputational damage. In the USA for example, legislation around alcohol sales is a nightmare. Every state has its own laws. There are different rules depending upon the alcohol content of a product, how much wine a person can buy each year and whether the wine is coming direct from a producer or via an intermediary. Amazon now sells wine online in the USA, but it took Amazon three attempts, and more than ten years, to get its wine offering right, mainly due to the complexities of state legislation.[10] If Booles were operating its online wine service in the USA then at the very least it would be prudent for the marketing team to use the following override rules, regardless of the scores people received:

- Do not target people aged under 21 (Legal age at which alcohol can be purchased).
- Do not target people living in states with laws that prohibit selling alcohol via mail order/online (for example Mississippi, Utah and Kentucky).

Often override rules like these are applied automatically to generate a decision – such as do not target people who live in certain states. Sometimes, however, override rules are required to divert cases to a manual process, to be dealt with by a human expert. In European countries, EU data protection legislation[11] gives individuals the right to have any automated decision made about them reviewed by an individual. This right is rarely taken up, but organizations that use automated decision making (based on predictive analytics, rule sets or any other mechanism) must have a process in place to refer these cases for manual processing when they arise.

In some industries, such as financial services, insurance and retailing, there can be dozens or even hundreds of override rules applied in conjunction with an organization's predictive models. Some examples of the override rules that are very commonly applied by credit card and loan providers include:

- Never offer credit to recent bankrupts.
- Always accept credit applications from "VIPS" (e.g. directors of the company and their spouses)
- Never decline credit applications from members of staff without first sending it for manual review.
- Always decline credit applications from people with an annual income below $20,000.
- Don't grant credit to people aged under 18 or over 75.
- If an individual has applied for credit within the last 30 days and their application was declined, then decline them again, even if their credit score has improved.[12]
- If a customer already has one of the company's products (e.g. a mortgage) and they are applying for another product (e.g. a loan) then refer them for manual review if they only just fail the cut-off score for this product.

Having a few override rules to deal with a small percentage of exception cases will not have a significant detrimental effect on the benefits case for a predictive model. In fact, some override rules add value to the decision-making process if they are based on wider domain knowledge of a human expert. A good example of this is the first override rule in the aforementioned list – the rule to never offer credit to recent bankrupts even if their credit score is very high. The reason for this is that predictive models are only as good as the data used to build them. If the data used to build them (the development sample) contains no information about certain segments of the population, then the model won't be very good at predicting the behavior of people in those segments. Most lending institutions always have, and always will, decline credit applications from people who have recently been made bankrupt. Consequently, the development samples used to construct their predictive models don't contain any bankruptcy cases that were granted loans, and hence bankruptcy won't contribute to the credit score people receive. Experienced credit professionals on the other hand, know that recent bankruptcy is a very strong predictor of creditworthiness and one should not extend credit to such people without a very good reason for doing so. Consequently, implementing an override rule to decline all recent bankrupts, regardless of their score, leads to better decisions than ones based on the score alone. Having said this, in general, the more override rules you have the less value your models will add to what you do. This is because fewer cases get dealt with on the basis of the prediction that you have made about people's actual behavior. In the extreme, if your decision-making processes are driven entirely by deterministic rules that take precedence over what your predictive models tell you, then this means one of two things:

1. Predictive analytics is not the right approach to be taking for that type of problem.
2. You need to change your process to enable predictive analytics to be applied as it should be.

With regard to the first point, it's important to remember that predictive analytics is only useful if you don't know what someone is going to do – it's about increasing the level of certainty that you have about an unknown event or behavior. If there is only one possible outcome; i.e. the behavior is a certainty, then you don't need a predictive model to tell you this.

A great example of this was when I worked with an organization that wanted to use predictive analytics to identify people who were highly likely to have made mistakes when completing a form about their financial circumstances, such as their income and the value of their assets (property value, share assets and cash savings). The idea was that these people would then be contacted to confirm that the information that they had provided was right, or correct it if it was wrong. However, it turned out that the organization actually had access to an independent source of this data via a third party. All that it needed to do was compare the values provided on the form with the third-party information and target those where the values differed significantly. There was no need for a predictive model

chapter 3

Analytics, Organization and Culture

As someone from the UK, the sort of things that come to mind when someone mentions culture are Shakespeare, Buckingham Palace, Mark E. Smith, the King James Bible, Mary Poppins, hobbits, Charles Dickens, *Withnail and I*, Henry Moore and the League of Gentlemen. These are all quintessentially British things that form part of Britain's rich and diverse cultural heritage, but a society's culture is far more than just the music, art and literature it creates. In its broadest anthropological sense, culture is about the full spectrum of behaviors and activities that individuals within a society undertake. The arts are a part of this, but culture covers the way people do things, how the pecking order is determined, what is acceptable behavior, what is not acceptable and what is taboo.

When discussing organizational culture, what we are talking about isn't dissimilar to the culture of a society. Organizational culture is about how people behave in that organization; what their values and motivations are, what is the right way to do something and what is unacceptable. Each organization can be viewed as its own little society distinguished from other organizations by the individual patterns of behavior adopted by those that work for that organization. In some organizations the culture demands that men wear a tie, in others it's an oddity to see anyone not in jeans. In some organizations (particularly in Japan, the USA and UK) working more hours than you are contractually obliged is widely expected. In France, Spain and Greece few work more than they have to.

Sometimes cultural norms are enforced by rules which, if you break them, lead to disciplinary action. In other situations cultural norms are dictated by "custom and practice" – things are done that way because they have always been done that way. When someone joins an organization they adopt the culture or they risk being ostracized by their colleagues – "When in Rome do as the Romans." It's also important to realize that just like society, organizations also have sub-cultures. The debt management department has a very different ethos for doing things than the marketing department. The London office has one way of working, the Paris office another.

Cultures are not created spontaneously. They develop over time. Some reach a steady state that hardly alters from one year to the next; others are constantly evolving. Sometimes cultural change is slow and natural, sometimes change is much more dramatic and traumatic, driven by a new management team or new legislation. Organizational cultures are complex, and it's not always easy to foresee what might happen if you try and change one bit of the culture – there are often knock on effects and unforeseen consequences.

If you want to change your organization's culture then this is something that takes time and effort, and there is usually more than one change strategy that can be adopted. One option is to work from within, making small incremental changes that accumulate over time to move things towards where you want them to be. The other option is to overhaul things quickly in one go and then move on to the next challenge. When we apply this to predictive analytics, the question is whether you start by using analytics on a piecemeal basis to improve a business process over here, then another one over there and so on, or come up with some sort of master plan to transform the business in a strategic and integrated way. Each approach has its benefits and drawbacks, and how you approach cultural change will depend on a number of factors, including the appetite for change, how the powers within the organization view change and the level of acceptance or resistance to new ways of working within the workforce at large.

3.1   Embedded analytics

Every now and again I attend industry conferences on predictive analytics. One thing I have noticed is that people often talk a lot about the need for an "analytics culture." What they usually mean is building up a good team of highly skilled (and highly paid) data scientists who will somehow manage to do all the things that need to get done with regard to analytics.

Establishing a good team of data scientists is part of an analytics culture, but there are a lot of other issues that need to be addressed if an organization is to be successful in evolving a culture that accepts automated decision making based around predictive analytics. The other things tend to get forgotten about or left in the background. The mantra seems to be: "If you have good data scientists then the rest will follow...." Personally, I don't like the phrase "analytics culture." To me it implies that analytics is the be all and end all that overshadows all the other facets of an organization. Consequently, I prefer the term "embedded analytics" when discussing predictive analytics in the context of organizational culture.

When we talk about having an "analytics culture" or "embedded analytics," what do we mean? An organization's attitude to data, analytics and automated decision making, driven by predictive analytics, all contribute to an organization's culture. If an organization is going to successfully embed analytics it needs to:

- Have the commitment of senior managers who support the use of analytics.
- Have a middle management that has a good understanding of the benefits and limitations of data and analytical methods. To put it another way, they should know what is possible using predictive analytics and what is not.
- Value data. Data is a commodity to be utilized, not something to sit idly in a corporate data warehouse.
- Be "data driven." Decisions are evidence-based, drawn from what the data is telling you, rather than hunches and "gut feel." This does not mean human judgment is unimportant, but that were the data supports or refutes a particular view or idea you give it due consideration.
- Maintain a skilled analytical team that has the capacity to reach out to the business and communicate analytical solutions in ways that they understand.
- Have buy-in from the "front line" who must act upon the decisions that result from using predictive analytics to predict behaviors.

So what this means is that the use of automated data driven decision making, using predictive analytics, must be accepted across the organization. Everyone, from senior managers to front line staff, needs to be comfortable acting upon the decisions being made by the organization's automated decision-making systems. For senior managers this means being willing to "walk the walk" and take responsibility for analytics systems within their command structure, rather than just talking about the benefits that analytics may bring and delegating the responsibility downwards.

In some industries all these things are already in place. Predictive analytics is accepted and fully embedded by all the major players in that industry – it's part of their culture. In banking and mass market direct marketing, predictive models drive almost all customer interactions, and most people are pretty blasé about automated decision making. Predictive models have been used for decades and everyone takes them for granted – there is no debate about their value and no opposition to their use. These days there is even a regulatory requirement to use predictive models to calculate how much capital a banking organization needs to hold to cover any unexpected losses that it might incur.[1] So, if you are someone who works for a bank or marketing company, you may be wondering why there is a whole chapter on embedding analytics and organizational culture! The reason is that in other industries predictive analytics is not as well established, and almost everywhere where predictive analytics has been attempted there has been strong resistance to its use.

Usually the battle to use predictive analytics has been won, but not always. Sometimes efforts to implement predictive analytics have failed, even though on paper the advantages to the organization are clear. At a recent seminar I attended,[2] a delegate from a European Tax Authority described their six-year struggle to get a predictive model implemented. Front line staff had resisted it from the outset and it was only after several failures and a lot of effort that the model had finally been put to use. Yet, in other tax authorities, such as Australia, the USA and UK, predictive models have been used for years and the case for predictive analytics in tax collection has been proved many times over.

One of the most important factors in successfully embedding analytics in your culture is to have good change people: that is, people who know how to create the space to allow the data scientists to do their thing and who can get the rest of the organization to buy in to the solutions that they deliver. These people don't need to be data scientists themselves, but rather have the expertise to make things happen and be able to take the rest of the organization along with them. In this respect, getting a good quality analytics team in place is probably the last thing on the list that you need to get right, not the first. You need to create the appetite for predictive analytics before you start recruiting people to do it. I've seen more than one case where whole teams of data scientists have been brought into organizations, have developed all sorts of predictive models that had the potential to deliver real business benefits, only for the solutions never to see the light of day because of a lack of understanding and buy-in from all sorts of people.

3.2   Learning from failure

As I say, I attend industry conferences now and again. Another thing that I tend to notice at these events is that every presentation is a showcase for the organization and the individual presenter. Almost everyone has a success story to tell (as do I!), which is great, but you rarely hear about the failures. This is very different from the academic approach. Failures and "non-results" are seen as just as valuable in extending human knowledge as successes and positive outcomes.[3] As most management books will tell you, failure is a key part of learning, and as someone once said, "It takes a wise man to learn from their mistakes, but an even wiser man to learn from the mistakes of others."[4] Therefore, in the next part of this chapter I'm going to give some examples of failure – in particular, projects where predictive analytics has been tried, but failed to be implemented or never delivered any business benefit. Hopefully you can learn from these tales of woe and avoid such mistakes yourself.

Personally, I'm not embarrassed to say that I was involved in some of these projects, but others might be. Therefore, while all the cases are based on real-life projects, I'm not going to disclose the names of the organizations or individuals involved. I have also changed/modified a few things that might identify those individuals/organizations.

3.3 A lack of motivation

The first tale I am going to tell is about a medium-sized regionally based finance company that worked with retail outlets to provide finance for furniture and home appliances – sofas, washing machines, TVs and such like. For several years the company, a long-established and quite traditional institution, had been gradually losing market share to the large national players in the industry. They had finally realized that they needed to do something different if the decline wasn't to become terminal. The guys in the credit department had heard a lot about predictive analytics (credit scoring in particular) and were interested in developing their own credit scoring models. After some enthusiastic meetings with various board members, Geoff, who worked in the credit department and was responsible for the company's credit risk policy, got funding to take the idea forward. So Geoff called us in and we discussed what he wanted. We then dutifully wrote everything up as a set of business requirements, presented the requirements back to the senior management team and got their sign-off on the project. Four months later, after a lot of time spent gathering and analyzing the company's data, with update meetings and check points at key stages of the project, we delivered two credit scoring models – one for assessing new customer applications and one for assessing additional finance for existing customers. The models were presented to the board, the company received a full report describing the end-to-end process, plus detailed instructions about how the model worked and the logic required to calculate the scores. Geoff was happy, his managers seemed happy and therefore we were happy.

A couple of months later I met up with Geoff at an industry conference and asked him how things were going with the models.

> We haven't got them implemented yet – it's going to cost a lot more than we thought to re-engineer our mortgage and loan process to include the models. We've also got quite a few other things on at the moment, so we were thinking about pushing the implementation back to next year when we have more IT resource available.

I met Geoff again at the same conference the following year, and the models still weren't implemented. This time the problem was the new operations director – who had not been around when the models had originally been developed. He was opposed to changes to working practices and staff reductions that would invariably result from using the models to assess risk, rather than human underwriters. As Geoff noted rather bitterly, this apparent concern for his staff was probably because losing staff would mean a smaller empire for him to manage.

It was a couple of years before I met up with Geoff again. The models had never been implemented, and when he had moved to work for another organization the project had simply sunk without trace. Later that year the company, whose fortunes had never improved much, was taken over by one of the bigger players in the market. As a result, 90% of head office staff had been laid off. The company was now little more than a brand name, with all its customer accounts managed by the parent company's operations center.

One lesson I learnt from this story is that building a predictive model is often the easiest part of a predictive analytics project. At the end of the day, a predictive model on its own is worth little more than the paper it's written on. Redesign of business processes and getting a model implemented is far harder, more time-consuming and more costly. If you are implementing predictive analytics for the first time then you need to be willing to invest in the IT and process changes required to deliver automated decision making. This is a mistake that many organizations new to predictive analytics have made and one which most consultancies have capitalized upon. These days, suppliers of predictive analytics solutions prefer to offer full end-to-end solution management for a customer, with ongoing license fees and running costs, rather than just building a few models for a customer as a one-off.

There is a second broader lesson. The finance company didn't disappear just because it didn't implement credit scoring (although that may have been a factor). The real reason is that it hadn't thought through and committed to the process of change, and therefore never made the strategic decisions that were required to survive in what was a rapidly evolving financial services sector at the time.

A final factor in the company's failure was one of ownership and responsibility for the models. Geoff, with all good intent, had been the project champion. Yet he was not senior enough to get it pushed through – the Operations Director was able to get the project sidelined with little resistance from his peers. One thing that should have been done was to make the model development and implementation the responsibility of someone higher up the management chain, ideally a suitable member of the board – so giving them an incentive to see the project completed successfully.

3.4   A slight misunderstanding

The next story I want to share with you is about a large European organization, with a diverse customer base comprising many millions of individuals and businesses. A key task for the organization was screening its existing customers to identify fraudsters. These were people making false claims for cash payouts or those not paying the organization as much as they should. The organization operated a three-stage approach to fraud identification, as shown in Figure 3.1

The first stage involved generating "short lists." A short list was generated automatically using a few simple rules that identified customers who had one or two risk factors associated with them. Each short list contained anything from a few dozen to a few thousand "high-risk" cases. For example, the "multiple payouts" list contained details of people who had received many payouts in a short period of time. Another list contained details of customers who had only made small payments, but whose assets and income suggested that they should be paying a lot more for the services that they received. The rules used to generate the short lists were reviewed each year and changes made based on what the fraud experts thought the significant risk indicators were.

In the second stage, the case review team (comprising a number of fraud experts) reviewed the shortlists and identified the cases that seemed most suitable for

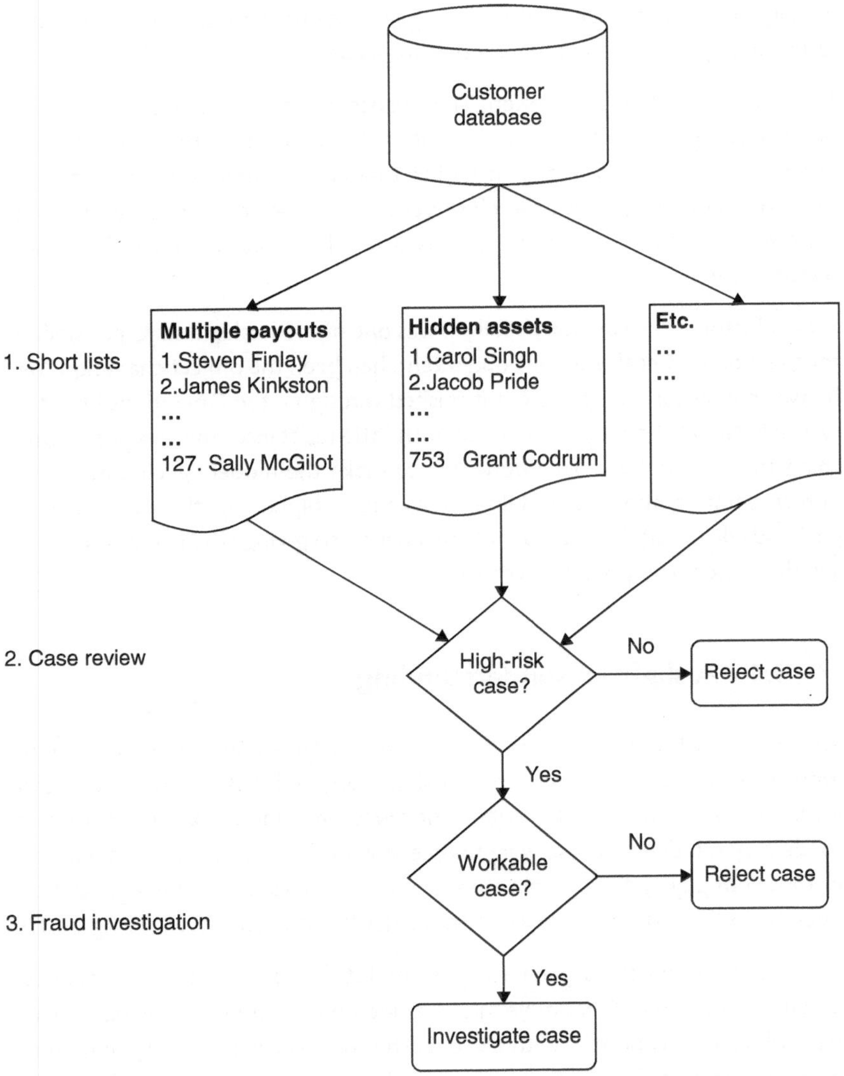

```
                          ┌─────────────┐
                          │  Customer   │
                          │  database   │
                          └─────────────┘
```

Multiple payouts   **Hidden assets**   **Etc.**

1. Short lists   1.Steven Finlay   1.Carol Singh   …
                 2.James Kinkston   2.Jacob Pride   …
                 …
                 …   …
                 127. Sally McGilot   753   Grant Codrum

2. Case review   High-risk case?   → No →   Reject case

                                    Yes

3. Fraud investigation   Workable case?   → No →   Reject case

                                    Yes

                               Investigate case

FIGURE 3.1 **The fraud process**

investigation. Typically, only about 30–40% of cases on any given list would be selected as "high-risk" and the rest discarded. The case review team then handed over the cases that they had selected to the fraud investigation unit.

In the third and final stage, the investigation unit considered the cases it had been provided with and carried out a detailed investigation of a case if they felt the case warranted it. An investigation involved reviewing the paperwork, asking the customer questions, making a site visit if that was appropriate and referring cases to the authorities if large scale-organized fraud was suspected. However, it was not unusual for the investigation unit to reject cases that the review team had provided.

The reasons for rejection were many, and occasionally caused some friction between the two teams, but rejections usually had something to do with the financial targets that the investigation unit had been set. The targets were based on the amount of money the unit recovered. A case might clearly be fraudulent, but if there were no money or assets to recover then there was no point investigating it.[5] So cases would be rejected if the customer under investigation had just died, had moved overseas or had no assets to speak of. Some cases would be rejected because investigators thought that they were difficult cases, and would therefore take a lot of time and effort to resolve. Their time would be better spent chasing up a larger number of simple, quickly resolving cases, increasing the overall amount of recoveries.

I also heard of a few situations in which cases were rejected for quite bizarre reasons. One example was where an investigator only took a case requiring a site visit if it happened to lie between the office and where the investigator's children went to school, so that he could then pick up his kids rather than heading back to the office. If the case was off the beaten track it would end up in the rejection pile.

As you can probably imagine, the organization wanted to make its fraud investigation process more efficient. The organization decided that it would use predictive analytics to replace the short lists and case review team. A predictive model would be used to predict the probability of fraud for each and every one of the organization's many millions of customers. So this was going to be a classification model, predicting whether or not someone was likely to be a fraudster. Customers would then be ranked by score, and those at the top of the list would be passed directly to the case investigation team – making the case review team redundant.

The organization hired a consultancy firm to build the model, and on paper the performance of the model was pretty good. A blind test based on a historic sample of fraud data showed that the model was able to identify segments of the customer base where the risk of fraud was many times the average and much better than most of the existing short lists. On the basis of these results the organization launched a pilot program. The highest scoring 0.025% of customers were sent directly to fraud investigators without any short lists or case review being undertaken.

The pilot was a disaster. The proportion of cases rejected by the investigation unit shot up from just under 10% to more than 60%. Why? There were several reasons:

- There were a higher proportion of complex cases. Yes, there was good money to be found, but they involved too much time and effort for the investigators to feel it worthwhile pursuing them.
- Some of the cases were located at bogus addresses, making investigation impossible.
- A lot of cases related to bankrupts without any assets.
- The model generated cases "without obvious risk." Yes the cases were highly likely to be frauds, but close examination of the case did not reveal any one specific thing to account for the risk. This made it difficult to open a line of enquiry with the customer. The investigators were used to beginning their customer dialog with something along the lines of "We are investigating you because...."[6]

With regard to this final point, one of the things the review team did was to select cases because there was something that clearly didn't stack up and pass this information on to the investigator to use as leverage when investigating a case. For example, the customer claimed to have no income, yet was living in an area where the average property sold for more than $1m. Another example was where a customer had made a claim for their spouse but was unmarried. In summary, the organization had failed to realize the full value added by the review team. They didn't just identify cases that were fraudulent, but also considered the "workability" and value of the cases, i.e. those cases where the investigation team had a reasonable chance of finding the fraudster and recovering money from them.

To try to salvage the situation it was decided to reintroduce the review team to the process. To all intents and purposes, the score was now going to be used to generate a new type of short list. However, this simply moved the problem upstream. The cases handed to the investigation unit were fine, but the review team were only selecting around 15% of cases on the list – less than half the number they were selecting before. This was despite all the cases on the list technically being high risk. This meant that the review team had become less efficient and hence more costly. So the second pilot was also deemed a failure.

In a last-ditch attempt to get some usage from the model, the organization decided that it would continue to use its existing short lists, but attach the model score to the short lists, together with the variables that featured in the model, and then see if it improved the quality of the case selection process, the idea being that the extra information would help the review team make better and faster decisions. However, this was exactly the opposite of what happened. The review team actually took longer to review cases when the scores were provided, and there was a tendency to decline simple workable cases if the score indicated fraud. This was seen as the last straw and the project was abandoned. So what had gone wrong this time? Three things were identified:

1. The extra information that had been provided meant that there was more information to consider, and hence the review process took longer.
2. The reviewers didn't really understand what the score meant. Yes, they knew it meant that a case was risky, but not why it was risky. So they didn't make much use of it.
3. The review team knew what it meant for their jobs if the model proved to be a success. Some staff adopted a deliberate policy of not selecting cases if the score was high in an attempt to discredit the model's ability to predict fraud.

The biggest failure of this project was a lack of understanding of the existing process; what happened at each stage and why. Another fatal mistake was to give the scores to the review team and let them decide how to use it. As a rule, people don't accept that simple models, such as scorecards and decisions trees (or more complex models) can do a better job of prediction than they can, particularly if they are experts in the field. What you tend to find is that humans will go with their own judgment, rather than trust the model score every time – so the model becomes useless. Models aren't perfect and get it wrong sometimes, just like people, but not quite as often. Those that oppose the use of predictive analytics will always pick out the cases where the model got it wrong, rather than acknowledging where the model got it right.

There are a number of approaches that could have been taken which would have improved the chances of the project being a success. One option would be to change the model. The model that was constructed predicted the probability of a fraud, but what would have improved matters is if it had been designed to predict the value of funds recovered.[7] Override rules could also have been applied to remove many of the "unworkable" cases, such as bankrupts and the deceased. Another option would have been to leave the existing process more or less unchanged, but use the model to remove low-risk cases from the existing lists before they got to the review stage; i.e. rather than using the model to identify high-risk cases, it could have been used to automatically deselect low-risk cases. This would allow the human assessors to concentrate on the best cases in each list, increasing the selection rate and improving efficiency, but without any changes or disruption to the existing fraud investigation process.

3.5   Predictive, but not precise

One organization I worked with required customers to complete a complex online form that asked them to provide all sorts of information about their employment, income, expenditure and lifestyle over a number of years. Analysis showed that around 20% of people made serious errors when completing the form and this caused a lot of problems further down the line that were expensive to resolve.

A proposal was put forward that maybe we could use the information on the form, plus additional information we held about customers, to predict in real time who had made an error the moment the form was completed. The idea was that we could send customers an e-mail telling them they had probably made an error within seconds of them submitting their form. As well as informing them of their mistake, the e-mail would contain a link back to the form, giving them the opportunity to amend and resubmit it before it was processed further.

The initial reaction from the business was positive. Preliminary analysis showed that it was possible to use predictive analytics to identify a group where the error rate was around 75%. If we could target the 75% group and reduce the error rate by even a couple of percentage points, this would translate into significant savings when processing customer information later on. However, the problem with this approach soon became apparent when we started thinking about how the model would be used operationally. The scenario was something like this:

1. We use the model to identify those who have a 75% or greater chance of making an error on their form.
2. We send these customers an e-mail telling them that they have made an error, and allow them to modify and resubmit their form.
3. The customer follows the link in the e-mail.
4. The customer amends their form and resubmits it.

OK – that sounds reasonable, but there were over 50 questions on the form. The model would only generate a single score that would predict if something on the form were wrong – it would not tell us anything about which of the 50 questions might be in error. So what is the customer going to do when they get an e-mail telling them that they may have completed the form incorrectly? They will ring the call center and ask: "What have I got wrong?" and our answer will be "We don't know!" The situation was made worse by the fact that the model was only 75% accurate. Twenty-five per cent of customers that we contact will actually have got things right, and all we are doing is encouraging them to change something that will increase the error rate. Consequently, the project was canned at the feasibility stage.

The project team dic play with the idea of building models to predict specific errors, i.e. a model to predict if the customer had made an error on question one, another model to predict error on question two and so on. However, we didn't follow this up for two reasons. One was complexity. Managing just one or two models is a non-trivial task. Having anything up to 50 models, all predicting different things just to check one form, didn't make business sense. Second, the predictive ability of the individual models would probably have been too low to have been of practical use. Why? Because predictive models tend to work best when they are focused on predicting clear, obvious things. When it comes to more nuanced outcomes it's much harder to build models that deliver the same level of accuracy. In banking, predicting whether or not someone will default on a credit agreement within the next six months is a much easier task than predicting how many days in arrears they will be in six months' time. Likewise, models that predict the month in which an insurance claim will occur are less accurate than a model that simply predicts whether or not an insurance claim will occur by a given date. That's not to say it's impossible to build usable models to predict things like this, but the task is more difficult and the results are often less accurate.

Although the idea for the form-checking model was not taken forward, the concept of using models to predict customer errors didn't die completely. Instead of trying to identify problems at the time when the form was completed, the team instead looked at a model that was applied prior to the customer completing the form. The action that resulted from someone getting a high error score was to tailor the cover note and instructions that were provided with the form. When the model predicted a high likelihood of a mistake, greater emphasis was placed on text suggesting ways that the customer could seek help when completing the form. In addition, a list of the "top five most common errors" was provided. This proved to be a successful approach to reducing the error rate.

3.6   Great expectations

My final tale concerns a UK mortgage provider. The bank had been using predictive analytics for many years to do all sorts of things, from marketing through to debt collection, and was a strong advocate of data-driven analytics and automated decision making.

In the mortgage market, a key driver of profitability is foreclosure (repossession of the mortgaged property). Mortgage providers use predictive analytics to estimate the likelihood of foreclosure. These predictions are then used to make decisions about how to manage customer relationships, such as whether or not to extend further borrowing (low probability of foreclosure combined with high equity) or when to initiate foreclose action, if an account is in arrears and it is unlikely that the customer will recover their ability to make repayments. These models are also very important for capital calculations that determine what assets a bank needs to hold in reserve to cover any unexpected loss events that might occur.

For the majority of customers struggling to meet their mortgage repayments, the path to foreclosure is, sadly, pretty straightforward. They get behind with their repayments and various attempts are made to reschedule the debt, offer payment holidays and so forth to give the customer time to pay,[8] but if the arrears continue to accrue then the mortgage provider will foreclose.[9] However, a significant minority of foreclosures are voluntary.[10] The customer hands back the keys to their property before any legal proceedings are undertaken. The customer has concluded that there is no realistic way for them to make their repayments. Therefore, they decide to end things as quickly as possible and make a fresh start, rather than enduring months of being chased by the debt recovery department before repossession inevitably occurs. What is interesting about voluntary foreclosure is that there are often no arrears, or the customer has only missed one or two payments. It occurs out of the blue with little or no warning.

Voluntary foreclosure is a bit of a headache for mortgage providers. This is because homes that have been repossessed typically sell for well below market value. If someone is in arrears and foreclosure is a distinct possibility, then the mortgage company will account for the potential shortfall in their impairment charges (provisions for bad debts). However, when a home owner hands the keys to the property back to the lender without warning, the lender is faced with a sudden and unexpected loss, often tens (and sometimes hundreds) of thousands of dollars per property.

What the bank wanted to do was use predictive analytics to forecast voluntary foreclosure within the next three months. Where the model predicted that voluntary foreclosure was likely, the bank would contact the customer and ask them to come in to their local branch for a face-to-face "financial review." On the basis of the review the bank would do one of three things:

1. If the customer was struggling, but the mortgage was potentially salvageable, then agree an action plan with the customer before the arrears began. This might include rescheduling the debt, putting the customer in touch with debt charities, or even just providing a bit of moral support.
2. If the situation looked hopeless, then help the customer to sell their home while they were still in it. This would ensure that the full market price of the property would be realized, benefiting both the borrower and the lender.[11]
3. Nothing, if it appeared that the customer's finances were sound (i.e. the model got it wrong).

To meet the objective, the analytics team spent a couple of months gathering data and building what at first sight was a great model. Statistical measures of how well the model distinguished between voluntary and non-voluntary foreclosure were superb.[12] Using the model it was possible to identify a group of customers where the likelihood of voluntary foreclosure was 120 times greater than the average customer.

The eventual show-stopper for this project arose when it came to implementing the model. Yes, the model was very predictive in statistical terms, but what the business had lost sight of is that foreclosure is a rare thing and voluntary foreclosures even rarer. In any three-month period, the bank typically foreclosed on about one in every 2,000 of the mortgages on its books,[13] and of these, about a third were voluntary. So over a three-month period an average of one in every 6,000 customers chose to hand back the keys to their home. As discussed, the model could identify cases that were 120 times more likely to enter voluntary foreclosure than average. This equates to a customer group where one in 50 (6,000/120) were likely to hand their keys back.

The bank wanted to invite the people in this customer group to their local branch and review their situation, but for every 50 reviews undertaken, on average only one would be with someone who was likely to hand back their keys. The bank estimated that each review required about 2.5 hours to deal with.[14] Therefore, about 120 hours (three weeks) of staff time would be required to deal with just one case where voluntary repossession was likely. The cost of this action was far more than the bank was willing to accept and consequently the project was not taken forward.

The problem in this case was not an unwillingness to accept analytics and automated decision making. Nor was it any lack of organization willingness to use predictive analytics to drive the business. The problem was the opposite – management were too trusting of predictive analytics. The cultural view was that predictive analytics could be applied successfully to almost any problem and would therefore be applicable in this case too. The business failed to differentiate between a good statistical model and a good business (practical) model.

At the post-project debrief, it was agreed that no business objectives had been set around the abilities of the model beyond a somewhat subjective statement that it needed to be good at predicting voluntary repossession. This was probably because the bank had never had to think much about this before. It used predictive analytics to predict fraud and credit risk within its marketing and debt management functions, and the models were always useful, so nobody had thought to question this one. With hindsight, it was agreed that the mortgage review process would only have been viable if the proportion of voluntary repossessions in the review population was at least one in four – more than 12 times "purer" than the 1 in 50 group identified by the model. To put it another way, to be useful the model would have had to be been able to identify customers that were 1,250 (6,000/4) times more likely than the average customer to voluntarily foreclose. At the outset of the project, the project team should have asked about the operational use of the model and drawn out this requirement – possibly doing some "quick and dirty" feasibility work to get some ball park figures to see if such a predictive model was viable before going further. To any experienced model developer, the requirement for an uplift of 1,250 times average should immediately have set alarm bells ringing. Models that are this good in credit markets (and most other application areas) are almost unheard of.

3.7   Understanding cultural resistance to predictive analytics

Pulling together the themes from the previous case studies, you can represent the issues that you need to overcome in terms of the culture cycle, as shown in Figure 3.2.

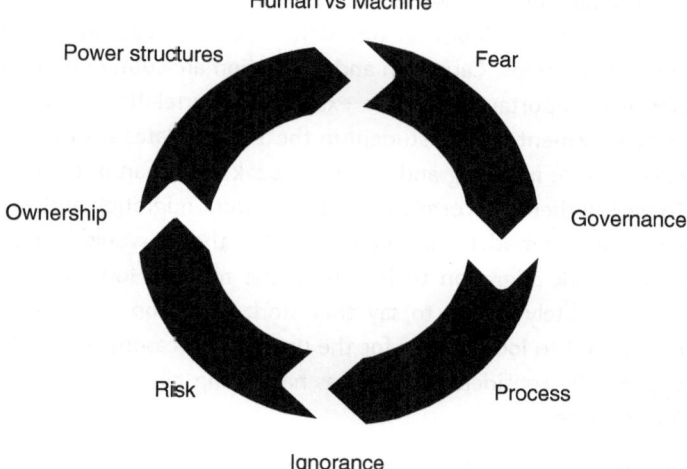

FIGURE 3.2 The culture cycle

Let's start at the top of the cycle, with Human vs Machine judgment and then move clockwise to address each of the other points. There is lots of evidence that an expert in a field won't trust predictions made by predictive models over their own judgment. They find it hard to accept that a simple scorecard or decision tree can make as good or better predictions than they can, with their many years of experience. If a model-based decision disagrees with their own view they will go with their own judgment every time. So if you just give experts a score and then give them complete discretion over how it's used, the model won't provide any benefits over the current decision-making process.

People also display bias when evaluating predictive models. If you present people with examples of where a model was used to make decisions and the eventual outcomes from those decisions, they tend to focus on cases where the model got it wrong rather than where it got it right. In particular, there is a tendency to pick out cases where the model was wrong, but a human got it right. I've never yet come across an example of a predictive model that gets it right every time, but in almost every case predictive models get it wrong less often than human decision makers. However, you only need a few cases like this for people to be convinced that the model doesn't do a good job, resulting in a lack of faith in the model's abilities.

One of the key causes of conflict between automated and human-based decision making is that models and people make predictions in different ways. In the world of human judgment and decision making, we look for reasons as to why something occurs – we are driven by *causation*. Predictive analytics on the other hand simply looks for correlations between data items, not the cause of the relationships. Causation and correlation are subtly different, but the difference is important. A classic example of correlation, which I first heard as a management science student in the 1990s, relates storks and birth rates. Storks migrate in spring and there is a peak in human births in spring as well! So we say there is a correlation between stork migration and births. If you have a predictive model which predicts births, then it would be perfectly reasonable for stork migration to feature in the model. However, you and I know it's completely wrong to say that stork migration *causes* births. A human would tend to look deeper for the underlying reasons why births are higher in spring, by considering what was happening nine months before in the previous summer.

One consequence of this difference in approach between correlation and causation-based reasoning is that if you show a decision tree, scorecard or other type of predictive model to a human decision maker, their assessment of how good the model is may not be based on how well the model predicts. Instead, they will assess the model based on how well the model mirrors their own decision-making criteria. People make statements such as "*The model can't be any good because no points are allocated for gender, and we all know that gender is a critical factor when it comes to predicting....*" That may be true, but it may be that other variables in the model act as a proxy for gender. Nursery school teachers and cocktail drinkers tend to be female, engineers and beer drinkers tend to be male. If occupation and preferred drink are in the model then there may be no need to have gender as well, even if there is clear evidence that gender is predictive of behavior – the other two variables take gender into account. Alternatively, the human decision maker may be making an incorrect statement. They think gender is important for predicting behavior, but in reality it's not. The predictive analytics approach on the other hand is *data-driven*. Gender will only feature in the model if there is a proven correlation with the predicted behavior *and* having gender in the model improves the model's predictive ability over and above the other data items that already feature in the model. Alternatively, you may find that a human decision maker can't understand why occupation and preferred drink are important in the model because they would never consider those things important when making their own judgments.

It can be difficult to completely address these sorts of concerns, but there are ways to mitigate them. You can show people the evidence – present some simple graphs or tables showing how different data items in the model correlate with behavior based on what has been observed in the past. Forty per cent of nursery school teachers exhibit the behavior, but only 5% of engineers. Therefore there is a clear correlation between occupation and behavior. Alternatively, it is possible to construct a model so that it gives greater prominence to data items that human decision makers prefer and less prominence to other data items, with no significant loss in the model's predictive ability.[15] It can also be prudent to exclude certain data items altogether, even if those data items are predictive. A small loss in predictive ability is seen as a reasonable price to pay for the model being accepted by business users.

Fear is the next item in the culture cycle. Many people fear the introduction of predictive analytics, and with good reason. Predictive analytics means efficiency, and efficiency usually means fewer people, deskilling and lower pay. Consequently, those that feel threatened by predictive analytics will take defensive action to protect their positions – which is quite understandable. For predictive analytics to be successful you need to put in place the process and governance structures to ensure that the decisions made using the model score are acted upon. To put it another way, a decision is only as good as the action that's taken as a result of that decision.

Two of the biggest reasons why predictive models fail to add business benefit are poor management information and weak governance. The decision-making process is not monitored to see if the correct actions are being taken and insufficient controls are put in place to ensure that score-based decisions are executed correctly. A very common failing is for managers to hold a training session and then send a few memos round to people telling them not to override the model. Some people forget, others ignore what they have been told and some will actively seek to discredit the score-based decisions being made. So in simple terms you need to be tough if you are going to make predictive analytics work for you. The three things you need to do are:

1. Remove the opportunity to override score-based decisions.[16]
2. Put in place the management information systems to be able identify where overriding occurs and by whom.
3. Take firm action against those that don't comply.

The easy option is to threaten to fire people if they don't comply. It's true that one aspect of good management is being able to take decisive action when it's required, but that doesn't mean you must set out to put people's backs up, frighten them or force them to do things unless there is no other option. A good communications strategy, explaining why the organization is changing, why new processes are required and so on, can go a long way to mitigating bad feeling. A few chats with the right people over coffee can work wonders. Many people don't like change, particularly if they feel threatened by it, but they are more likely to respect the change process if you can persuade them that using predictive analytics is the right thing to do. In the long term that's a much better option in terms of good employee relationships, staff turnover, productivity and so on than forcing unwanted change upon them.

Ignorance is another issue that needs to be addressed. Most people don't have the slightest idea of what a predictive model is, how one is developed, their benefits and limitations or how they are used – it's a magical black box. So you need to up-skill your people to a sufficient level. I don't mean that you need to teach people the technical skills about how to build a predictive model, but the right people should know what a predictive model looks like, how a score is calculated and how it is used; i.e. the sort of things we covered in Chapters 1 and 2.

An approach that I follow is to deliver a tailored education and training strategy targeted just at the key people who need to know about predictive analytics. For middle managers I deliver a one-day classroom-based training course. This contains no formulas or equations, but provides people with examples of what scorecards and decision trees look like, how to make a score-based decision, the benefits and limitations of predictive analytics and so forth. For senior managers I have a one-hour version that I deliver as 1:1 sessions with a particular focus on opportunities (benefits), strengths, limitations and risks around predictive analytics.[17]

Sometimes people think that everyone in the organization needs to know about predictive analytics. In the past I have been asked to develop educational materials suitable for thousands of front line staff working in branches and call centers. However, my view is that this is a mistake.

People only need to know about the models and how they are used if they are going to be involved in their development and deployment. For front line staff acting on the decisions made on the basis of model scores, this does not mean they need to know all the ins and outs, what variables feature in the model or even that a predictive model is being used. The most important thing is that they are provided with the information and resources to do their job. If predictive analytics results in different types of cases being worked, then staff may need training in how to deal with these new cases. You may need to supply additional information in support of the decision-making process, but you don't need to mention predictive analytics to do that. A good example of this is the second case study, discussed earlier in the chapter, with the fraud investigators. As well as removing the "unworkable" cases such as bankrupts and deceased, it would also have made sense to develop a set of indicators/flags that highlighted any particular concerns or issues, which could then be used as the lead into a case. The fact that a predictive model was used to select the cases isn't important if the fraud investigators have the information they need to work the case optimally.

Like anything new, implementing predictive analytics has risks, and predictive analytics is not always a success. You need to have an appetite for risk if you are going to have a chance of success. A lot of organizations talk about a "no blame culture" or "acceptance of failure," but in reality failure is rarely rewarded. This leads us on to issues of ownership. One of the biggest risks for any type of project is a lack of ownership. I don't mean not having a project manager, but rather someone with sufficient seniority in the organization to be able to push things through when they need pushing, who can authorize expenditure and resources, and who takes personal responsibility for championing the project. In a risk-averse organization that punishes failure, you will find an unwillingness amongst senior managers to take on something culturally challenging like predictive analytics because of the personal risks that the project represents. So you need to find ways to overcome that.

Finally, there is the issue of power. If implementing predictive analytics means that you change process and culture, hire new people and fire others, then you will find yourself rubbing up against some of the powers within the organization. Some people will want to prevent predictive analytics impinging on their patch; others see predictive analytics as an opportunity and want to steal your fire. So you need to be able to negotiate between interested parties to see that the organization's goals (and your goals), are delivered, rather serving the interests of specific individuals or groups within the organization. Consequently it's important to get buy-in from the right people and understand who will be influential in deciding how predictive analytics develops within the organization. Some questions to consider:

- Who will affected by predictive analytics – who are the stakeholders?
- Who will gain if predictive analytics is implemented?
- Who will lose if predictive analytics is implemented?
- Who is going to champion the use of predictive analytics?
- Who are the gatekeepers, without whom change cannot be achieved?

In very crude terms, if you are going to "Win" at predictive analytics then those that have something to gain have to overcome those that have something to lose. You need to consider which side has the greatest influence and authority as they are the ones likely to win. What this means is that predictive analytics may naturally be owned by one function, but it may be politically expedient for some other part of the business to take responsibility for it, if you think there is greater chance of it being pushed through if it sits with that center of power. However, what you absolutely don't want to do is hand responsibility for the implementation of predictive analytics to someone who is going to lose out from it.

The final group in the above list are the gatekeepers. These are the enablers who have indirect power and influence, but not necessarily any direct interest in whether the project is a dramatic triumph or dismal failure. Often, these are people in corporate functions such as HR, IT, Finance and so on, that don't have anything to do directly with delivering products or services, but oil the wheels of the organization, ensuring that the infrastructure required to run the business is delivered and maintained. If you need to acquire resources, develop new IT, hire people and so on, then you need to keep these people onside.

3.8   The impact of predictive analytics

The context within which a model is developed and deployed within an organization is very important in terms of its impact upon that organization and how it operates. Some models can serve their purpose with little or no impact on the way the organization works – the model is a tool that a particular department decides to use to help it work better, and no one else in the organization needs to know about it. Maybe you have a work queue and the model is simply used to sort the queue in priority order.

In other cases creating the necessary environment to use automated decision making can have consequences across the organization, even if the model is built and owned by a single department within that organization. If a model is going to be used to drive automated pricing of insurance products, then if you get the model wrong it will have a very serious impact on customer recruitment and bottom line profitability.

To help me assess the likely impact of introducing predictive analytics to an organization and how much disruption it might cause, I like to consider the following things:

1. **The scope of the model.** A model with narrow scope addresses a single specific problem, usually within one well-defined business area. There are

few implications for the wider business. In contrast, a model with wide scope affects the whole way that an organization approaches its business processes.

2. **The risks that the model presents.** If a part of the business is going to be dependent upon the model, and won't be able to function without it, then the model creates an operational risk that needs to be managed. If model failure is more of an inconvenience rather than life-threatening, then risk mitigation is a much lower priority.

3. **Manual or automated actions**. After a decision has been made, it needs to be acted upon. Will the required action be undertaken automatically, or will an individual be responsible for doing it?

There are no hard and fast rules, but a narrow scope model is often a "one shot" model used tactically. It is developed, used for a particular job and then discarded. A targeting model, such as the one used by Booles in the case study from Chapter 2, is a narrow scope model. The model was used a couple of times over the course of a year with the specific purpose of identifying potential customers, and that was it. The way the model was used had very little impact on other areas of the business outside of the marketing department,[18] and the marketing team will probably build a new set of models next year. Another example of a narrow scope model is an R&D (proof of concept) model. A model is built to see if something is feasible, or how much value a model might provide. Possibly the model is tested "live" by putting a few cases through the model to see what happens, but the model is not used in earnest.

At the other end of the spectrum, wide scope models have strategic impacts, often on an ongoing basis. The model is not used just once or twice, but many times – often in real time. A bank's credit scoring models will be used every single day to assess credit card and loan applications, and the models have impacts far beyond the credit department. Change the way the model is used (the cut-off) and it will impact marketing,[19] funding, capital requirements[20] and operations management – all of the core functions of a banking operation.

The second point on the list is about risk. What will happen if you get the model wrong? Who will be hurt? How much money will be lost? Whose reputation will be damaged and who is likely to lose their job? If the impact is large, then you are much more likely to encounter resistance to the implementation of the model. At one end of the risk spectrum, predictive models developed as R&D carry little risk and it really doesn't matter what the outcomes are as long as it answers the research question that was proposed. At the other end of the risk spectrum models are used for things like insurance rating, credit assessment and identifying tax evaders. Get these models wrong and millions, perhaps billions, of dollars could be at stake as well as reputational damage to those organizations and the individuals that work for them.

The final component of my assessment is the mechanism for dealing with people after decisions have been made about them. If the resultant action is automated, then you probably don't have as much to worry about from a cultural perspective. If all that happens as a result of your decision making is that a promotional text message is sent to people via your CRM system, then nobody is really concerned about why specific individuals are targeted and not others. If on the other hand, your predictive model is used to identify people who should have a limb amputated, then the highly trained surgeon who is going to do the job has to have faith in the model. If nothing else, the Hippocratic Oath prohibits a doctor from doing harm, and if the surgeon has a different view of whether a limb is salvageable or not, then they will ignore the model and follow their instinct. We saw similar behavior in the second case study, with the fraud investigation team. If a case didn't match an investigator's expectation of a fraud, then they rejected the case, regardless of the statistical evidence that pursuing the case was the right thing to do.

So in summary: Models that are deployed locally, where resulting actions are dealt with automatically and where there is little risk associated with model usage, are the easiest to deploy. Conversely, models that have strategic, organization-wide implications, are business critical and impact downstream manual processes, require the greatest cultural change and buy-in before they are accepted and put to use. If a model falls into this category, then you need to make sure you have everyone on board and committed before you start, otherwise there is a significant risk of the project unraveling when it comes time to put your models to use.

3.9   Combining model-based predictions and human judgment

When a human expert and a predictive model are presented with the same information about people, the predictive model will, on average, make better (more accurate) predictions than a human expert. However, human expertise can still add value to the decision-making process, and for some types of decision making it can be a mistake to try to remove all human involvement.

In the previous chapter we discussed the use of business (override) rules. These are rules put forward by business experts, which are implemented alongside a predictive model and take precedence over the model score in some situations: for example, never target under 18s with promotional offers for tobacco products, even though the model score predicts that they are highly likely to take up the offer. The reason why these business rules add value is because they draw on wider human experience to provide additional information that goes beyond what can be inferred from the raw data alone.

The design of automated decision-making systems, based on a mixture of data-driven predictive models and expert derived business/override rules, is one way that human judgment and predictive models can be combined successfully. Another way of combining these two methods of prediction is at the level of an individual case. Human judgment can add value to model-based predictions if the human expert can bring some extra information to the table that isn't available to the predictive model. In tax collection, for example, local tax inspectors often have access to information about small businesses on their patch that is not maintained in centralized computer based tax systems. For restaurants they can do a "drive by" and see how full it is, then have a look at the menu prices and come up with a quick estimate of what the restaurant's turnover should be. If their estimate differs substantially from what the restaurant declared, then this is a good reason to investigate further. Likewise, a doctor can get a lot of extra information from how a patient looks, what they say and how they say it, which can help improve diagnosis over a model based just on physical evidence, such as blood tests, scans and previous medical history.

As mentioned earlier in the chapter, if you present a person with a score from a predictive model and then let them make their own judgment, they will usually defer to their gut feel. The score becomes useless. If you are going to combine human and automated prediction at the case level, you need to do it in the right way. Here are some options for doing this:

1. **Use the model to make decisions in the majority of cases.** Marginal cases which just pass or fail the cut-off score are referred to a human expert to make the final decision.
2. **Use the model as a prioritization tool.** All decisions are made by a human expert, but the best (highest scoring) cases are dealt with first.
3. **Distill the population**. The poorest (lowest-scoring) cases are removed. Human experts then review the rest and make the final decision about each case.

Which of these options to take forward, if any, depends on several factors. This includes business, cultural, legal and reputational issues. In particular, the level of human involvement in individual decisions will be dependent upon the volume and value of those decisions (and the cost of getting it wrong). If we are talking about making life and death decisions about thousands of patients that a model predicts do or don't have a particular disease, then you will probably want to deploy a good deal of expert opinion in conjunction with the model – so we are talking about using the model to do things like prioritizing waiting lists (Option 2). However, if we are talking about a mass market organization sending out millions of promotional offers for a cheap consumer product, then there is probably no value at all in having marketing professionals deciding who should or should not be sent an offer – it's simply not practical or cost-effective, and there is no comeback from consumers, politicians or the police if you get it wrong. The best option is to automate the entire process without any human involvement in individual decisions.

4
chapter

The Value of Data

Data is a key ingredient in the predictive analytics process. You can't do predictive analytics without it. In general, more data leads to more accurate models, but not all data is useful. In fact, most data is completely useless and it is very much a case of diminishing returns as more data becomes available. Doubling the amount of data you hold about your customers will not double the predictive accuracy of your models. Likewise, using twice as many customer records to construct a model won't double the accuracy either. There is also a cost/benefit case to consider, in terms of both the number of data items you consider for inclusion in a model and the number of individual records that are used to construct models.

On the plus side, the cost of storing, processing and analyzing data has reduced dramatically in recent years, which makes it cost-effective to use bigger samples and consider more data items than ever before. In the early 1990s a gigabyte was something that could only be processed by mainframes and supercomputers. There was no practical way to analyze a gigabyte of data outside of a few universities and specialist research centers. Today, a multi-terabyte[1] hard drive for your laptop/PC/server costs less than $100. In today's world it doesn't cost much to store a lot. Consequently, the attitude taken by many organizations is to store as much as they can and then worry about finding the useful bits later.

One argument against a Big Data "store it all" philosophy is that in the past, when storage was so much more expensive than it is today, people spent a lot more time thinking about data. They made very conscientious and well-thought-out decisions about what data would be useful to them in running their businesses, and they made sure that they captured that data. Data was only kept if it was useful. If it wasn't useful it was discarded. If one wants to take a somewhat negative stance on Big Data, you could say it's a lazy approach, akin to sifting the garbage that people didn't think was very useful first time around in the hope of finding a few additional scraps of information.

I don't hold to this view, but there is some truth in it. Data storage is cheap these days, but it's not free. As we discussed in Chapter 1, if you already have good data systems in place, extending this to a Big Data view which incorporates machine data, unstructured and external data such as text files, server logs, telematics, call transcripts and so on may give you benefits, but these benefits could be more marginal than you had hoped (or the hype from vendors suggested). This is one of the dangers of jumping into solution mode with this type of Big Data. It's often difficult to establish how much benefit you are going to get before you do it, and you could end up spending a lot of money and not achieve very much.

If you are planning to go down the route of storing and exploring all the unstructured (and other) data that you can lay your hands on, make sure you do some feasibility work that establishes the value of the data, have clear go/ no go decision points in your project plan and be willing to ditch the project if the results of the feasibility study do not suggest that the data adds sufficient value. Alternatively, if you are planning to source a project externally, you could suggest that the supplier works on a profit share basis rather than a fixed fee, taking a percentage of any added benefit that their solution provides. In this way, if the benefits are smaller than expected you won't be out of pocket.

4.1   What type of data is predictive of behavior?

Regardless of what behavior is being predicted – whether it's someone's propensity to go on a second date with you or the likelihood of having a heart attack in the next 12 months, most predictive models use no more than 20–30 data items (variables) to generate their predictions, and some considerably less than that. The scorecard model for loan applications in Chapter 1 contained only eight data items and the decision tree from Chapter 2 was based on just six. Personally, I have never come across a model that needed to include more than about 30 data items. That's not to say you can't have a model with more, but your model is probably more complex than it needs to be. This is important, because although you many have considered thousands of different bits of data on the way to building your models, requiring you to trawl through many terabytes of data, you actually only need a very small part of all that data when it comes to implementing those models: creating scores and making decisions.

The big question is: Which 20–30 data items are the important ones? Given the explosion in personal data there are often thousands, or even tens of thousands, of data items one might consider. So there is a lot of work to be done to find just the important data items that you need for your model. Thankfully, with the capabilities of modern computer systems, sifting a few thousand data items and finding the handful that are most correlated to the behavior you want to predict doesn't take a huge amount of time or effort.

What complicates matters is that there is almost always other data that could be acquired or generated in addition to what is already available and this extra data doesn't always come for free. The most prominent sources of external data are credit reference agencies and database marketing firms, which hold thousands of data items about most of the adult population. They slice and dice, combine and refine the data they hold in all sorts of different ways, which is then served up for different types of predictive application. Organizations then pick and choose the data items they want to purchase from the menu, but until they do some analysis of the data they can't say with any certainty whether that data adds value over and above the data they already have.

Another consideration is derived data. To develop a good predictive model one doesn't just use the raw data that's available. The most predictive data items are often derived from two or more other pieces of data. The loan amount requested and the borrower's income might both be predictive of future loan default, but industry experience is that the ratio of loan amount to income is far more predictive than the two component pieces of data on their own.

So how do you decide what data is likely to be important? One option is to rely on "brute force" to examine every possible alternative, gathering all the data you can and throwing it at cutting edge hardware/software. However, even the most modern massively parallel computer systems will struggle to consider all possible variations and combinations of the data. It can be more productive to start by applying some common sense, using the opinions of subject matter experts to consider what types of data are likely to be most predictive and focus on that first, before getting into other, more abstruse, data types. When considering the data available for a modeling project, I classify it into the types shown in Table 4.1.

Table 4.1 is *source agnostic*. At this point we are not talking about where the data comes from or what format it's in. Instead, we are talking about the

Table 4.1   Data types

Data type	Description
Primary behaviors	Information about a past behavior that is of the same type as the behavior you want to predict. If you build a model to predict burglary, the fact that someone has committed burglary before will be a strong predictor that they will do it again.
Secondary behaviors	Information about past behaviors which are similar (but different) to the predicted behavior. Littering and non-payment of parking fines aren't major crimes, but they indicate a propensity to commit worse offences. This is why the police pull over cars with broken tail lights; there are probably other crimes and misdemeanors to be discovered.
Tertiary behaviors	Information about past behavior that does not have an obvious or direct connection with the behavior being predicted. For example, the fact that someone sold a lot of items on eBay recently may give some bearing on their propensity to burgle.
Geo-demographic	Details about a person's state of being, rather than what they have done in the past. For example, their age, blood pressure, weight, hobbies, what car they drive, their income and occupation.
Associate data	Information about other people, someone has a connection too. This includes the associate's behaviors (primary, secondary and tertiary) and their geo-demographics. If someone's spouse has a history of burglary, there is a greater propensity for them to commit burglary as well.
Sentiments (feelings and opinions)	A person's attitude to things; their likes/dislikes, approvals/disapprovals. If someone answers a crime survey by saying they think it's OK to steal, or they write a derogatory blog about law enforcement, then they may be more likely to engage in criminal activity than the average individual.
Network	This is information about the nature of the connections between a person and their associates. If someone has 100 friends and 99 of them are burglars, then they are likely to be a burglar too.

relevance of data items to predict the behavior of interest. In Table 4.1, I've used burglary as the example of what we want to predict, but the same data classification can be applied to all sorts of prediction problems, from insurance and consumer credit through to book-buying behavior and health.

In Table 4.1, I've used an immediate family member (spouse) as an example of an associated individual. However, there are many other types of association. A financial association exists if you jointly take out a loan with a friend, or enter into a business partnership with someone. Consequently, *their* previous behavior could be a good predictor of your future behavior. Many organizations consider the postcode/zip code in which you live, assuming that you share the same attributes as your neighbors and will therefore behave in a similar way to them. Your behaviors are also correlated with those of your friends, work colleagues and classmates. Even more tenuous links, such as belonging to the same organization, club or society, can provide some useful insights into your behavior, based on the behavior of other people who are members of those groups – even if you don't know these people personally.

There are no rules written in stone, but what I find time and time again, regardless of the type of behavior I am trying to predict, is that data about primary behavior is the most important by a considerable margin. It often accounts for 80–90% of the predictive ability of a model. If you've done it once you'll do it again. If your credit card company offers you a credit limit increase, this is mainly because of the (good) repayment behavior you have exhibited in the past 6–12 months when paying your credit card bill. If you only ever dated sporty types in the past, then you will probably date sporty types again in the future. If you buy frozen peas every week at the supermarket, then that's a pretty good predictor that you will buy frozen peas next week too.

After primary behaviors, information about secondary behaviors adds most to the accuracy of predictive models of human behavior. If you pay your mortgage and utility bills on time, then that's a good indicator of how you pay your credit cards. If you buy lots of frozen beans, sweet corn, carrots and so on, then you might also be good for frozen peas. I also find geo-demographic data and associate data (their primary and secondary behaviors) add significantly to many types of predictive model, and the closer the association the more useful the data: information about one's nearest and dearest (family, friends and business partners) is more important than information about people with other types of association.

In my experience, tertiary behaviors tend not to be very predictive. For example, what restaurants people visited, where they went on vacation and who they voted for will generally have only the weakest of associations with whether someone defaults on their credit card, buys frozen peas or dates sporty types. It's true that every so often someone will put forward a really interesting and surprising behavioral association that they have found (such as beer purchases are associated with diaper (nappy) sales[2] or that insurance claims can be predicted using information about credit usage[3]), but these associations are relatively rare, and where they exist they are fairly weak, i.e.

they provide only small incremental benefits to models constructed using information about primary and secondary behaviors and geo-demographics.

Sentiment data is interesting if you can get it, particularly if the sentiments that are expressed relate directly to the behavior you want to predict – in effect acting as a surrogate for primary and secondary behaviors where you don't know them. If someone writes a letter complaining to a store about the quality of the frozen peas they bought last week, then there's a good chance they won't buy frozen peas again in future. If they write a letter praising the quality of the peas, then the store can probably exclude that customer from future discount offers for peas and still get a sale. In the same way, if someone writes about a great date they had on a social network site, then that gives some insight into the type of person they like to date and so on. Sentiment data is one of those things that has really come to the fore in the Big Data world, but in practice is hard to do. In many ways, the science and application of sentiment analysis is still in its infancy, and consequently it is one of the least utilized types of data in predictive models at the time of writing.

With regard to network data, I have found its greatest value to be as an enabler, to establish who people's associates are, and it's the information about the associates that adds most value to the model. What tends to be less valuable (except in a few specialist cases)[4] is data about the structure of the network – for example, how many people someone is associated with, whether they are at the edge or in the center of a network and so on.

I find that thinking about data in this way can be very useful, particularly when time and resources are limited. My advice is to concentrate your efforts accordingly. It might be worth paying for, and investing a lot of time to obtain, data about primary and secondary behaviors and geo-demographic data, but the return that you see from other types of data is unlikely to be so significant, and therefore you need to think more carefully about the cost/benefit case for obtaining it.

4.2   Added value is what's important

When considering different types of data, it is very important to realize that data items are not completely independent of each other. If I know your date of birth, then I don't need to ask you your age because I can calculate it. Age and date of birth are said to be *perfectly correlated* – one can be determined from the other. If age/date of birth are predictive of behavior, then when

I build a predictive model, one or other of the variables will feature in the model, but the accuracy of the model will not be improved by including both.

Perfect correlation is rare: more often data is *partially correlated*. If I know the age of your spouse I can make a pretty good guess at your age. If your spouse is 35, then there is about a 90% chance that you are aged between 30 and 40, but there are a few exceptions, so the correlation isn't perfect. What this means is that I may not know your age, but if I know the age of your spouse I've pretty much got things pinned down. If I do subsequently find out your actual age then that will give more of an insight into your behavior, but not that much more.

What tends to occur is that most data is correlated with one or more other pieces of data to a greater or lesser extent. The presence of correlation means that when assessing how valuable a data item is, just because that data item is strongly predictive of behavior on its own does not mean that it adds value over and above the data that you already have. If the new data is highly correlated with existing data then it won't add much to the power of your predictions.

When assessing new data items you need to assess them in terms of added value, not as standalone items. To illustrate this, let's think about a situation where we are trying to predict whether the next car that someone buys is a Volvo. Imagine that there are two data sources available to construct a model:

- Car purchase history over the last ten years.
- Internet tracking data that contains details of the Internet sites people visited in the last few weeks.

Let's assume that the organization already has data about car purchase history[5] and this is stored in a nice well-structured format in its data warehouse. The organization is considering upgrading its IT systems and analytical capability to be able to process the additional Internet tracking data and carry out analysis of the web-pages that people visited. To evaluate the worth of the data the organization builds three different models. The first model is constructed using only existing data about car purchase history. This includes data items such as which types of car people bought in the past, the age of the cars, their mileage, how much they paid for them, how long they owned each car before replacing it and so on. For the sake of argument, let's say that this model has an accuracy of 40%, i.e. the model correctly identifies people whose next car is a Volvo four times out of ten.

They then build a second model using only data items derived from the tracking data. For example, using variables that represent the number of times each car brand was mentioned in the sites that customers visited, coupled with sentiment analysis to create variables indicating whether those car brands were discussed in a positive or negative way on those sites. The accuracy of this model is 30%. What happens if you now build a third model using both types of data together? A naïve expectation is that the model will have an accuracy of 70% i.e. the combined accuracy of the two individual models. In practice, what you are likely to find is that the accuracy of the combined model is only slightly better than the best single data source model – perhaps somewhere in the region of 42–43% in this example.

That's a benefit, but perhaps not as much as expected. Why does this happen? Think about it from a common sense perspective – if each of the individual models had an accuracy of say 60%, then that would mean the combined model would have an accuracy of 120%, which is impossible! You can't get it right more than 100% of the time. What is happening is that both sets of data are overlapping (correlated) to some extent. The Internet data tells you something extra, but a lot of what it's telling you can already be discerned from the purchase history data. One often hears figures thrown around like, "85% of customer data is unstructured"[6] but the important thing is the incremental value of that data. It doesn't matter how much unstructured data is available if it's not giving you something extra.

On this basis you might think that unstructured (Big) Data adds little value, and in some contexts you would be right. However, in many mass markets, where we are talking about millions of potential customers, a 1–2% improvement can translate into a very large financial benefit that justifies big IT spending. It is also the case that, when it comes to marketing applications, you often don't have much of the really good behavioral information about individuals or their associates. If this is the case, then geo-demographic data tends to dominate these types of models. However, even geo-demographic data can be thin on the ground. This is one reason why Big Data has proved so interesting to marketing people, and marketing applications are what you tend to hear most about when Big Data is being discussed. By trawling the right sites, distilling information from previous customer communications, tracking the movements of people through their phone and so forth, one can get some level of insight into people's behavior that would not otherwise be possible. The behavior of associated individuals found via (social) network analysis is also seen as an important proxy for an individual's potential behavior. If all your friends drive old bangers, then there is a good chance you drive one as well, and won't be very interested in buying a brand new car. If lots of people you work with have the latest smartphone, then you are probably a good prospect for one if you don't have one already, and so on.

4.3   Where does the data to build predictive models come from?

In Table 4.1 the different types of data that are useful for building predictive models were explored – primary and secondary behaviors, geo-demographics and so on. The reason for discussing this first, before thinking about data sources and the format in which data is provided, is that when we talk about the value of machine data, blogs, Internet tracking data, e-mails and so on, this type of data does not have any intrinsic value because of where it has come from. It has value because it is contains data about primary and secondary behaviors, geo-demographics and so on, which are the most important types of data for predicting future behavior – the challenge is to extract that information.

Table 4.2 captures the main sources of data that are available for building predictive models.

In Table 4.2, each data source is described in terms of how structured it is, from high to very low. Highly structured data is usually numeric (e.g. income, age, speed of travel) or categorical (e.g. residential status, country of origin, brand of car). If data fits nicely into a column on a spreadsheet then it probably has some structure.

Unstructured data, such as customer correspondence, images and sound, is more varied. It's also possible to have semi-structured data. If you allow people to enter their occupation as a string of text (rather than selecting it from a list), then what you won't get is a nice number or neat categories, but it will be fairly easy to identify the most common occupations, such as teacher, cleaner, builder, manager and so on. Likewise, names and postal addresses have a degree of structure, but can contain considerable variation in how they are presented.

The key difference between structured and unstructured data is the amount of work that needs to be done to get the data into a useable, structured format that is required to do predictive analytics, i.e. the data must be in either a numeric or categorical format. When data is already well structured it can be used more or less as it is, with only a little bit of "data cleaning" required. Unstructured data on the other hand (text and images in particular), requires

Table 4.2 Data sources

Data source	What data is provided?	Level of structure	Origin (internal or external)?
Customer transactions and account records	• Geo-demographics • Contact details • Purchase/usage history • Payment history	High	Internal
Customer relationship management (CRM) systems	• History of inbound and outbound customer contacts	High	Internal
Web-logs/ Clickstreams/ Page tagging*	• The path taken to get to a site (via tracking cookies) • The pages visited within a site • Time spent on each page • Your computer's IP address (hence your approximate geographical location)	High	Internal or external
Power dialers	• Date, time and length of phone conversations • Date and time of attempted, but non-connected calls • Recordings of calls and outcomes	Medium	Internal
Sensors	• Telematics (in-car tracking of how a car is being driven) • GPS (from mobile devices) • Biometrics (heart rate, blood pressure etc.)	High	Internal or external
Credit reports/ credit reference agencies**	• Geo-demographics, financial information, credit history, fraud indicators	High	External
Mailing lists	• Contact details and geo-demographics	High	External
Customer correspondence	• Letters, e-mails, texts and phone transcripts	Low	Internal
Cameras and imaging systems	• Pictures • Videos • Medical imaging (e.g. X-rays, scans)	Very low	Internal or external
Surveys	• Mixture of geo-demographic, behavioral and sentiment data	Depends on survey design	Internal or external
Web-sites/ pages	• Text, tables, pictures and video	Low	External
Social networks	• Geo-demographics • Sentiments and opinions, likes and dislikes	Medium	External
Tweets and blogs	• Text about what people are doing, how they are feeling, their thoughts and opinions, their likes and dislikes	Low	External

*Web logs provide information about webpages someone has requested to visit. Page tagging uses code within a webpage to collect information about the site visit.
**At one time the information on a credit report was pretty much restricted to previous credit usage, bad debts and bankruptcy information. These days credit reports also contain all sorts of additional geo-demographic, behavioral and lifestyle data.

a lot of processing to transform it into a structured format prior to predictive analytics being applied. For example, one thing you can do with textual data is to count the number of times that certain words appear, and it is the word counts that are then used in the predictive analytics process.

When we are talking about predicting the behavior of existing customers, the most important data source, by a considerable margin, is an organization's own (internal) IT systems, in particular customer transactions and accounts records, because this usually contains a large amount of information about primary behaviors. A typical retailer, bank, government agency and so on, would have gathered a significant amount of behavioral and geo-demographic information when they first entered into a relationship with an individual and will have detailed records about all of the transactions that have occurred with the individual since that relationship began.

The source of data is also important. If data is internally sourced then an organization has control over:

• The format of the data
• Data security/privacy
• Data quality
• Timeliness
• Update frequency

The same can't be said of some types of externally sourced data, which raises questions about the stability of the data over time. If the format, content or quality of the data can change without warning, then the predictions generated by the model are at risk of being corrupted.

When building predictive models, if an organization does not have much information then the first port of call should be an information conglomerate, such as a credit reference agency or database marketing company, who make a living out of gathering and selling data. These organizations have followed a Big Data philosophy for years, and have decades of experience of gathering all sorts of diverse data. They then undertake the hard work of transforming and filtering it to generate high-value structured data that is easy for their clients to utilize. The fact that these organizations are developing their own approaches to new Big Data sources, tailored to the needs of specific industry sectors, to some extent mitigates the need for individual organizations to invest in their own Big Data solutions. Why develop the infrastructure to trawl hundreds of millions of web pages to gather information about your customers when all you need to do is supply a list of names and addresses to a third party who will do all the hard work for you? If you don't even have the contact details of people you want to contact, then these organizations will happily provide you with lists of individuals that meet whatever requirements you specify.

4.4   The right data at the right time

You need good quality data to do predictive analytics, but you also need timely data. It's not enough to know everything about your customers as at today. You need to be able to recreate the customer journey through time. If you are going to get the best use out of predictive analytics then you need data at the following times:

1. Historic data from the past, which is used to build the model.[7]
2. Current data at the point where the model is going to be applied to generate new predictions.
3. Management Information (MI) to allow you to assess how well the model is working after it has been implemented.

As discussed in greater detail in Chapter 8, most predictive models are constructed using historical data from the past. Data about customers (predictor data) is taken from one point in time and their behavior observed sometime after that (outcome data). The predictive analytics process then generates a model by examining how the predictor data is related to the outcome data.

Let's think about a predictive model for a dating site that is used to match people together. The model score indicates how compatible you are with other people registered on the site; i.e. how likely you are to hit it off. You are then introduced to people who score the highest.

You can measure the success of a dating site in a number of ways. Success can mean a second date, or it can mean marriage and kids, but for argument's sake let's assume that a match is deemed a success if the couple are still together 12 months after being introduced. If it takes 12 months before you can decide if a match was a success, then to build your predictive model you need to gather predictor data about matches that were made at least 12 months ago, and link this to information about each couple's relationship status as at today, i.e. 12 months after they were introduced.

One problem with some IT systems, particularly operational systems that customer service staff use to support customer interactions, is that they only maintain a current view for some types of information. If that information changes then the original data is lost. If you want to do predictive analytics then you need to be able to maintain a historical perspective of your data. So for our dating model, if you want to include information such as how long someone has lived at their address, their income and so forth, then you need to ensure that you capture that information at the time when each introduction is made. If you rely on current data, then it's possible that this information will have changed, and therefore your model won't be as good as it could be.

For internal systems that organizations maintain themselves, the historic data problem is relatively trivial. Most organizations have systems that record customer events in such a way that the historic data position can be recreated easily. This drives both their business information (BI) systems and their predictive analytics. In simplistic terms they have a "time machine" – they can enter any date in the past and recreate a view of their data as it was at that time. However, the need to obtain a historical view of data for predictive analytics can be a problem for externally sourced data. In particular, it can be a limiting factor for some sources of Big Data, such as web pages, social network sites and so on. Even if a site is date stamped, what you can usually see is only the current version of the site, not what it contained historically. Even worse, what you sometimes find is that the date stamps on web pages don't exist or are not updated when their content is updated. This is mostly a problem for applications of predictive analytics that have a long maturity period. For example, for credit scoring and insurance rating the time between capturing predictor data and observing the associated behavior (outcome data) can be one to two years or more. This is much less of a problem for direct marketing applications, where the outcome period is rarely more than a few days or weeks, and it is reasonable to assume that the data on the website won't have changed much in the intervening period.

The data used to construct predictive models comes from the past. However, when it comes to using the model, the data typically comes from the present, and that data needs to be accessible at the point where the model is going to be applied. I remember hearing one story about a retailer that had carried out a detailed survey of several thousand of its customers (out of a customer base of a few million), gathering all sorts of information about them, their lifestyle and purchase behavior. They then used this to build a model that predicted whether someone was likely to be a high-spending or low-spending customer for product targeting purposes. The idea was that they would be targeted with offers for budget or premium versions of products, depending on their wider spending profile.

To make a prediction the model used existing information such as customer's income, where they live (and hence the size/value of their home), age, gender and so on, but also a lot of the more detailed lifestyle information gathered as part of the survey. This included data items such as the type of car someone drove, where they liked to go on vacation, if their kids went to state or private school and so on. From one perspective the model was very reasonable and predicted well. However, when it came to implement the model within the organization's customer relationship management (CRM) system it quickly became apparent that most of the information used by the model wasn't available for most customers. Yes, they had some of the information required, such as age, income and address, but they didn't have all of the other stuff such as the type of car, children's education and so on.

To get around this problem, the obvious and naïve option would have been to ask customers the additional information at appropriate customer contact points, typically, in-store or online at the checkout. After answering the questions the model score could be calculated and an appropriate marketing offer sent to them, to be used the next time they made a purchase. However, this was deemed impractical. One reason was that it would have meant costly changes to the company's IT systems to prompt customers with the necessary questions and then store the results. Another issue was that if the questions were to be asked in-store, then staff would need some training, and governance structures would need to be put in place to ensure that they asked the right questions in the right way and recorded the results accurately.

The real killer though, was around the impact on customer service and the customer experience. If you think about a how a typical store (physical or online) interacts with its customers then it's all about providing a simple, quick and efficient service. How would you feel about being asked six or seven questions about your lifestyle at the checkout with a queue of people behind you? If you are trying to complete an order online, then being asked a whole load of additional lifestyle questions is a pain, and a real disincentive to completing the transaction. You may also have concerns over privacy and how the data is going to be used. Consequently, the company came to the conclusion that the negatives associated with asking customers the extra information required to calculate the model score outweighed any benefits from improved targeting of promotional offerings that the model might provide.

What the organization should have done was to identify what information it already had available about (all) of its customers, and then built its predictive models using only this data. This second model would not have been as predictive as the first because it was based on less data, but from a usability perspective it would have had a much greater chance of success.

Historic data is required to build a model, and the data items that feature within the model need to be available at the point where you wish to apply the model. The final type of data that you need for predictive analytics is management information about each decision that was made using the model. Whenever a score is used to make a decision you need to capture the score and the data items used to generate that score, as well as the decision(s) made on the basis of the score. Most importantly, you need to track (monitor) the outcome from each of those decisions so that you can compare the predictions made by the model against what actually happened. This process of monitoring is important because predictive models tend to deteriorate over time – their ability to predict gets worse as economic, market and social change occurs. The relationships that were found between the predictor data and the outcome data when the model was originally constructed no longer apply.

Sometimes the change is sudden, for example as the result of new legislation or the economy crashing/booming. Alternatively, a company may decide to reposition itself and target a different audience with different characteristics and behaviors, requiring different models. At other times it's more of a gradual trend as society or the economy evolves over time. So what one tends to see is a gradual decline in the quality of the decisions made on the basis of the model score. Consequently, predictive models have a finite lifespan. In some domain applications, such as healthcare and credit risk, it may take years for a predictive model to deteriorate to the point where it needs replacing with a newer, more up-to-date model. Marketing response models typically last for a few months, often being updated in line with seasonal trends and new product offerings. At the other end of the spectrum, models are rebuilt on a daily or more frequent basis, where markets are changing in real time, for example when products are competing against each other on price comparison websites and the features of those products are changing minute by minute.

4.5 How much data do I need to build a predictive model?

Sometimes models are constructed using data about all the individuals that an organization has information about. If you have the resources available then that's not a bad way to go. However, for organizations with hundreds of millions of customer records and thousands of data items to consider it's not practical to use all the data unless one has access to a supercomputer or has invested in modern, massively parallel/distributed computer architectures, such as Hadoop or HPCC.

Even if an organization does have cutting edge IT and analytical capability, it can still be cumbersome building models using data sets of this size. Consequently, when the population is very large (tens of millions of cases), common practice is to use a smaller more manageable sub-set (a sample) of data to build the model. The question for organizations in this position is: What sized sample should they use? The model development sample needs to be large enough to be representative of the full population and hence allow the model to capture the patterns of behavior within that population, but small enough to allow the data to be analyzed fully.

When I say "analyzed fully," what I don't mean is undertaking a single pass of the data that delivers a static view of the customer population and a single predictive model at the end of the process. Instead, I am talking about the ability to slice and dice the data in many different ways and explore many different modeling options interactively and quickly. It is possible to fully automate model development, but the best models are still developed through many cycles of analysis and model building guided by human expertise. If your samples are too large you risk losing this interactive capability.

At the other end of the spectrum, some organizations have very few examples of behavior to work with. Although Big Data is what gets the attention, there are far more people with "small data" that they want to understand and use. Perhaps I've developed a phone app that a few hundred people have purchased and I want to build a model to identify the wider population that might be interested in buying it. In situations like this, the question is: Is there enough data to build a predictive model?

These questions about sample size may seem straightforward, but like many things in life, the answers are not clear-cut. There are no definitive rules on sample size for building predictive models, but there are lots of rules of thumb, some case studies and a lot of experience to draw upon, and these can provide us with some useful guidelines.

To construct a predictive model, data is analyzed to determine what attributes predict people's behavior, such as the age range into which they fall, whether they are male or female, married or single, and so on. To determine if an attribute is important there need to be sufficient examples of people displaying the behavior you are trying to predict who have that attribute. Typically, one needs at least 30–50 examples of an attribute to be able to say whether it's an important predictor of behavior or not. If you want a really good model, then several hundred examples of each attribute is preferable. If you have 1,000 customers, but only four of them have the attribute "male," what you can say with certainty is that your population is biased towards women. What you can't say is whether men behave differently from women. Even if all the women exhibited the behavior and all the men did not, four cases are not enough to infer if gender is important for predicting that behavior.[8]

The 30–50 rule applies to just one attribute. With a predictive model there are many different attributes that contribute to the model score. What this means is you usually require at least several hundred examples of customer behavior to build a model. In the past I have been asked if I can build a predictive model based on as few as 50–60 cases, provided to me as a one page list in a word document. Predictive analytics is a large population tool and the answer in cases like this has been no – I can't build a model with such a small sample.[9] Or put in a rather more nuanced way, any model built with such a small sample will make predictions with little validity.

If we are talking about building a model to predict something like who is likely to buy a new phone app, a usable model can probably be built with information about 300–400 people who bought the app. What you also have to bear in mind is that predictive analytics works by contrasting the differences between behavior and non-behavior. This means that if you want to build a model to predict who is going to buy the app, the model development sample also needs to contain examples of people who didn't buy the app. If you have 300–400 non-buyers as well, then that gives a minimum of around 600–800 cases in total.

Around 600–800 cases, half of which display the behavior you want to predict and half of which do not, is a minimum. Samples of this size will enable you to produce a usable model, but a larger development sample will result in a better (more predictive) model. There are two factors that come into play as sample size increases:

1. **The predictive analytics process becomes more accurate.** For a scorecard type model this means that the points allocated to each attribute are closer to being optimal. With decision trees better splits are chosen, and possibly these are more refined (with more splits and more end nodes in the tree).
2. **It becomes possible to include rare attributes in the model.** Think back to our sample of 1,000, of which only four were men. If a larger

sample of say 40,000 had been taken, then there would have been 160 men. This is enough to determine whether gender is predictive of behavior.

Given the above points, it's not uncommon to use between 10,000 and 100,000 cases to build a predictive model. However, a key issue is the number of cases in the population at large. For classification models what you often find is that people aren't split 50/50 in terms of the proportion that do/do not display a given behavior. A large bank may have five million mortgages on its books, but only 10,000 cases where foreclosure occurred. We say that foreclosure is the "minority class" and non-foreclosure is the "majority class," and it's the minority class that is the limiting factor. One approach would be to take all 10,000 foreclosure cases and then randomly sample 10,000 non-foreclosure cases from the rest of the population.[10] Likewise, you may have a list of ten million people that you offered a product too, but only a couple of hundred thousand of them bought it. There is no point including millions of the majority class if you only have a few thousand minority class cases.

The other limiting factor is computer resource. Building predictive models takes a lot of computer power. The amount of computer power depends on three factors:

• The number of cases in the development sample.
• The number of predictor variables under consideration for inclusion in the model (as opposed to the number that feature in the final model).
• The type of model being constructed and the algorithm being used to construct it.

Doubling the number of cases in the sample tends to result in a doubling of the amount of computer power required to generate a model.[11] However, doubling the number of predictor variables that are being considered for inclusion in the model often follows a squared relationship. If it takes one minute to analyze 100 predictor variables, it will take four minutes to analyze 200, 16 minutes for 400 and more than an hour for 800.[12] The type of predictive model that you construct is also important. The scorecard and decision tree type models that we looked at in Chapters 1 and 2 are relatively quick to construct. More complex types of models such as neural networks and support vector machines (which are described in Chapter 6) require two to three orders of magnitude more computing power.

As a rough guide, once you have samples containing more than 2–300,000 records and about 1–2,000 variables[13] model building starts to become unwieldy using a single PC/Server.[14] This is not to say that you can't use bigger samples with a single high end PC, but when building predictive models the process benefits from a degree of agility. Before a final model is delivered, a good data scientist will build many different versions of the model, exploring different methods of model construction, changing the way the data is formatted, including some variables and excluding others, to get the very best model possible.

If you have very large datasets then your ability to undertaken experimentation is dramatically reduced – any benefits that you gain from the larger sample are offset against the reduction in exploration and experimentation. Consequently, it can make a lot of sense to take more than one development sample. First, a small sample is taken, containing just a few thousand cases. This is used to explore the data and generate dozens, or possibly hundreds, of different test models that are evaluated against each other. Once the best model-building options have been found, a final model is then constructed using the full data set.

Personally, I like to start with quite small samples, where I can generate models almost instantaneously, in just a few seconds, see what the model looks like and move on, refining my approach as I go. A great benefit of this approach is that very quickly you have a model and results that you can discuss with business users. I'll only consider larger samples after a good degree of exploration and managerial discussion, and I have developed something that is pretty close to the final model. Another one of my rules of thumb is that I am uncomfortable if it takes longer to generate the model than it takes me to get a cup of coffee, and if it looks like it's going to take more than two to three hours I will take action to reduce the sample size and/or the number of predictor variables under consideration.

Every few years a myth circulates in the predictive analytics community that there are huge benefits to be had from using all of the data available to you to build a model. Only a fool would consider building a predictive model using just a few thousand cases when many millions are available. Yes, bigger samples mean better models, but how much better? What's the cost/benefit case? How much more money will the company make from a model constructed using a hundred million records rather than 10,000? If the answer is only a fraction of a percent then you are probably going to get more benefit sticking with a sampling approach and doing a lot of different things, rather than investing heavily in the technology to be able to do full population modeling.

From my own experience, anecdotal evidence from talking with other data scientists and experiments on sample size based on consumer credit data,[15] it appears that once you get above sample sizes of about 10,000 then the benefits of larger samples become marginal.[16] Increasing your sample size from 10,000 to 100,000 is probably going to give you a 1–3% increase in the predictive ability of your models. Increasing your sample from 100,000 to 10 million is likely to result in a similar level of improvement. To put it another way, if a predictive model constructed using 10,000 cases is worth $100m per annum to an organization, then increasing the sample size by two or three orders of magnitude holds out a promise of another $2–6m.

Another way to think about the benefits is to consider what you can achieve using a decent desktop PC costing a few thousand dollars compared to using the largest supercomputer available to mankind. The extra storage capacity and processing power of the supercomputer will give perhaps a 1–2% uplift over and above what can be achieved using the desktop, if you have enough data to make this option worth considering (hundreds of millions of records, tens of thousands of predictor variables).

One thing to be aware of is that sometimes the benefits of taking very large samples are exaggerated due to poor methodology and validation when building models. A critical risk when building a predictive model is a problem called "over-fitting." A model is constructed which appears to predict very well based on how it performs on the development sample, but when you test the model using a fresh set of data that was not used to construct the model (a validation sample), the performance is much worse. An experienced model developer knows how to spot over-fitting and how to deal with it, delivering models that are as good as they can be given the available data. The over-fitting problem is less likely to occur if large samples are used.[17] What you sometimes see is an over-fitted model constructed using a small sample being compared to a model constructed using a large sample where over-fitting has not occurred. Consequently, the difference in accuracy between the two models is larger than it should be.

5

Ethics and Legislation

The way in which governments, corporations and other organizations gather, store and use data about people has been the subject of much debate in recent years. As organizations hold more data and increasingly rely on automated decision making to decide how to deal with people so have concerns grown about the impact this has on us as individuals and as a society.

People often talk about the dangers of data as a single problem with a focus on privacy, particularly in the light of reports about government agencies monitoring our private data at a very detailed level. However, there are actually three areas of concern when we are talking about the ethical use of personal data:

1. **Privacy and ownership.** Who decides what information an organization can hold about us and who owns this data?
2. **Security and anonymity.** The mechanisms in place to ensure that our data is not obtained by unauthorized third parties.
3. **Decision making.** What decisions do organizations take based on the data they hold about us and how are those decisions made? Are decisions based on human judgment or automated decision making?

In the sections that follow we will address each of these questions, but first, let's talk a bit about ethics.

5.1   A brief introduction to ethics

Ethics is a fascinating subject that has been much debated throughout human history and has led to many an argument over the dinner table. The famous Greek philosopher Aristotle (382–322 BCE) wrote a whole book about ethics more than 2,300 years ago.[1] The enduring interest in ethics is hardly surprising. Ethics is something that affects each of us personally every day, in our relationships with others and how we behave as a society. What makes ethics so complex is that, unlike physics or mathematics, there are few fundamental truths. What appears ethical from one standpoint can be considered entirely unethical from another. In terms of defining ethics as a subject, we could get into a long and in-depth debate, but a simple working definition that I shall adopt is ethics as: "The study of right and wrong." It's about the appropriateness of how we behave and what we do.

An important differentiator in ethical debate is whether you view things from a consequentialist or non-consequentialist perspective. A consequentialist judges things based on outcomes resulting from an action. The means justify the ends. It does not matter what you do if everything turns out all right in the end. Non-consequentialists, on the other hand, follow rules or principles, with less regard for the outcomes. It's more about the journey than the destination. To aid our understanding of these two ethical perspectives, have a think about the following question:

• Is it ethical to kill one person to save the lives of two or more others?

If your answer to this question is an unconditional yes, then this consequentialist view could be taken to imply that it is ethical to carry out a medical experiment which kills the test subjects as long as it leads to the saving of a greater number of lives in the long run. If your answer is a clear no, on the principle that it contravenes basic human rights, then this implies a non-consequentialist logic to your decision making.

This is a very simplistic example, and further arguments can be presented from both perspectives that give different answers to this question. A consequentialist could argue that killing test subjects in medical experiments causes significant harm to wider society, due to the fear and bad feeling induced by the chance of becoming a test subject and from the emotional harm incurred by the relatives and friends of the test subjects. In that case, the consequentialist could decide that the action is unethical using a wider measure of "overall public good," rather than just considering the consequences for the individuals directly involved. It might also be argued from a non-consequentialist viewpoint that the *principle* of saving many lives is right, and therefore, killing the test subjects is ethical.

Utilitarianism probably is the best known consequentialist model of ethical behavior and is based on the idea of "the greatest good for the greatest number." The formulation of utilitarianism is attributed to the British philosophers Jeremy Bentham (1748–1832) and John Stuart Mill (1806–73). They believed that an ethical action is one that maximizes happiness and pleasure within the population. The only unethical action is the opposite – that which leads to pain or unhappiness. Why maximize happiness and not some other objective? Because in Bentham's view it was the only thing desirable as an end in itself. All other things are only of value in so far as they lead to increased happiness.[2] However, some people believe there are things other than happiness and pleasure that can be considered to be intrinsically good (such as gross domestic product or life expectancy). Therefore, it is possible to be a consequentialist, but not a utilitarian.[3]

For utilitarianism to be practical, the amounts of happiness and pleasure caused by different actions need to be measurable. This is so that alternative actions can be compared and the best one chosen. One approach is to represent happiness and pleasure in terms of utility – allocating a numerical (often monetary) value to the desirability of each outcome. The outcome that yields the greatest overall utility is taken to be the most ethical. An alternative to utility is to measure the popularity of actions. In this case, the belief is that happiness and well-being can be represented in terms of popular approval. A hospital is performing well if people report that they are very satisfied with the treatment they received, even if that hospital's death rates are higher than average.

Kant's Ethical Theory[4] advances a non-consequentialist view of ethics based on duty and respect for others. Rather than asking what the impact of an action will be, Kant argued that an action is ethical if it is shows respect for others.[5] In other words, one should be motivated to act in good faith, out of a sense of duty as to what is right and proper, which Kant termed the Categorical Imperative. The fundamental idea captured by the categorical imperative is universality. Ethical behavior is something that every rational person would agree with. A great example is the problem of how to choose the best way of dividing a chocolate cake.[6] Everyone might want the whole cake for themselves, but the only consensus (and hence ethical) decision that can be arrived at by rational people is to divide the cake equally amongst everyone.[7]

Kant also presented a second formulation of the categorical imperative as a respect for individuals, which can be taken as complementary to the principle of universality. People should not be treated as objects, but as sentient beings with the right to their own life and opinions. People should never be used merely as a means to one's own ends. This idea is often expressed as the Golden Rule: "Do unto others as you would have done unto you." However, this implies that we should carry out actions such as giving all our money to complete strangers, because this is what we would want them to do for us.[8] Consequently, it is the negative version of the Golden Rule that is often most insightful: "Do not do unto others as we would not have done unto us." This places the emphasis on what we should do to avoid inflicting harm rather than maximizing the well-being of others – which Google expresses in its unofficial company motto: "Do no evil."

Religious teaching forms the basis of many people's lives, with God-given law forming the basis of the ethical framework within which they operate. The fundamental principle is that God is a better judge of right and wrong than we are. Perhaps the best known example of this type of law is the Ten Commandments, given by God to Moses: "Thou shalt not kill," "Thou shalt not steal," and so on. However, no system of religious laws deals with all possible situations that people encounter in their daily lives and the application of reason is always required to interpret and apply what is believed to be God's will in specific situations. There are no Holy Scriptures about genetically modified crops or the use of nuclear weapons, but it is still possible to formulate a theologically derived moral stance on these issues.

Aristotle believed that for everything in nature there was a right and proper purpose – a natural law to which it should conform.[9] This principle also applied to human beings and human activities. Your ultimate purpose is to be fulfilled in your life. Aristotle believed that one way for a life to be fulfilled was to undertake social responsibilities, such as being a good parent or a conscientious worker. However, fulfillment also means to act as a rational being, guided by reason, so that we control our emotions, live disciplined lives and do not give into sudden temptation to do foolish or extravagant things.

Aristotle called these principles "virtues," and concluded that there were inherently virtuous modes of behavior. Over the next two millennia this idea evolved into the concept of natural rights, or human rights, as we know them today. There are certain things that an individual has a given right to do, and certain things an individual cannot be denied, that should be beyond the power of any individual, government or other power to grant or deny. The role of government is simply to accept and protect the rights of its citizens. One of the key figures in formulating these ideas in a modern setting was John Locke (1632–1704) who promoted the concept of the right to life, liberty and property for all. These ideas were later incorporated into the American Declaration of Independence in 1776 and the French Declaration of the Rights of Man in 1789.[10] Today the list of generally accepted human rights has grown to include concepts such as the rights to free speech, free association, religious freedom and choice of sexual orientation.

One question that naturally arises is: Where do these rights come from? The traditional view was that they were "God given." Some modern philosophers find this a difficult position to adopt. An alternative view, formulated by John Rawls (1921–2002) is that we should move beyond the idea that society merely protects the rights of individuals, to one where we define rights as those that should be granted by a just society.[11] In many ways the concept of human rights can be considered a natural complement to Kantian ideas of duty and respect for individuals. So for example it could be argued that workers have a basic right to receive a minimum wage for their labor, while employers have a duty to provide fair wages.

5.2 Ethics in practice

Putting ethical theories into practice can be difficult. Arguably, all ethical frameworks can be found lacking when applied to some real-world situations. Pol Pot in Cambodia and Stalin in Soviet Russia both applied what can be described as extreme utilitarian ideals to government policy during the 20th century, putting the welfare of the state before that of the individual citizens that comprised the state. In both cases millions of people were deliberately killed as a result of their policies. The misinterpretation of religious teachings has been used as the excuse for wars, much cruelty and many acts of terrorism throughout history. It can also be argued that to follow a Kantian or human rights philosophy entirely can lead to equally poor decisions, and in the extreme can conceivably lead to massive catastrophes that could otherwise be avoided. This is because someone does what they believe should be done, not what avoids some ultimate disaster – a justification used in the 2000s by the US and UK governments to deny basic human rights to "suspected terrorists" at Guantanamo Bay and Belmarsh Prison respectively.[12]

Another problem is that ethical theory is easy to apply in certain clear-cut cases, but many issues are extremely complex or the full facts are unknown or disputed. People may also have different value systems, particularly if they have different cultural or religious perspectives. Individuals with a North American or Western Europe heritage may consider it unethical to offer a payment to a government official to speed up government bureaucracy. In countries such as Kenya and India,[13] what westerners would consider a bribe is a widely accepted practice engaged in at all levels of society. In some cases things won't get done at all unless a suitable payment is offered.

When people hold opposing views it can be very difficult to form a consensus as to whether a particular action is right or wrong, moral or immoral. Well-known issues such as abortion, euthanasia, animal rights, same-sex marriage and capital punishment all fall into this category, with it being possible to argue a different but convincing case from many different perspectives. In life, people naturally adopt ethical frameworks that encompass both consequentialist and non-consequentialist perspectives to generate their own individual views of what constitutes good behavior, and no two individuals will have exactly the same viewpoint on every subject. However, the one assertion that can be said to apply across all ethical frameworks is the notion that ethics carries with it the idea of something more important than the individual. An ethical action is one which the perpetrator can defend in terms of more than self-interest.[14] To act ethically, one must at the very least consider the impact of one's actions on others.

5.3   The relevance of ethics in a Big Data world

Is ethics relevant to a discussion about predictive analytics and (Big) Data? If you work for a commercial organization owned by its shareholders, with the aim of maximizing shareholder return, then this is not necessarily a stupid question. Making money is what commercial organizations do and what shareholders expect. I could argue that for an employee of a commercial organization to do anything other than trying to maximize the profitability of their employer is unethical because it acts against the interest of the shareholders to whom the employee has a responsibility in return for the wages they receive.[15] This includes doing charitable and other "good works" if they use the company's time or resources at the expense of making money – that's not what they are being paid to do. Where controls are needed to protect people against unethical action, it is the role of government to legislate on behalf of these people. Therefore, from this perspective, any action taken in a commercial setting is ethical as long as it complies with the law of the land. Legal compliance is a good thing because it is less costly than the legal fees, fines, imprisonment, loss of sales from bad publicity and so on, that arises from acting illegally.

If we lived in an ideal world, where the law is the embodiment of ethical behavior, then a "just follow the law" attitude to ethics would be an easy case to argue. However, secular laws are at best generalizations of ethical codes of conduct, representing a particular view of what constitutes good behavior. Often there is an overlap between what is legal and what is ethical, but it doesn't follow that just because something is legal it's also ethical. This is what we mean when we talk about someone following the letter of the law rather than the spirit of the law. A great example of this is UK Alcohol laws. In the UK it is legal to give (but not sell) alcohol to a five-year-old child. However, I don't know anyone that would think that it would be ethical to do so.

With regard to business practices, there are many examples of organizations using the letter of the law while ignoring the spirit of the law. Often this has resulted in short-term benefits, but when the practices these organizations employ becomes public knowledge, the long-term reputational damage and loss of customer confidence has been very significant. At one time banking was seen as a staid but honest profession. As a banker one took pride in helping people look after their money and supporting the local economy. Bankers were trusted and relatively well paid compared to their peers. After deregulation in the USA and UK in the 1980s, banking adopted a very short-term profit-orientated view, resulting in the bonus culture and all that went with it, cumulating in the financial crisis of 2007/2008. As a consequence, bankers are less trusted and less respected than they once were.

Another great example is tax avoidance. In 2013, there was a huge amount of publicity in the UK when it emerged that Starbucks had paid virtually no UK tax for years. Starbucks was accused of using complex tax loopholes to funnel UK profits to other countries. What made matters worse was when Starbucks responded to public criticism by offering to make a token tax payment as a gesture of "goodwill." The result was the opposite of what Starbucks expected, with even more vitriol from the public. Executives failed to understand that the amount of tax you pay is determined by law and enforced by the revenue service. Tax is not optional, and the UK public didn't like the arrogance implied by Starbucks' goodwill gesture. However, one has to remember that Starbucks wasn't acting illegally. It paid its accountants to limit its tax liability lawfully, and that's what they did – if they had turned around and said: "Well, you could have paid $100m million less tax last year, but we decided it would be nicer to pay more," then they would have been sacked, and possibly sued.

5.4 Privacy and data ownership

Who has the right to hold and use data about you? One (utilitarian) view is that data about you is a resource to which anyone can subscribe – information about your state of being, what you think, what you say, and how you behave is there to be harvested by whoever has a use for it. Having as much of your personal data as possible is a good thing, because it means that organizations can tailor their products and services to your requirements. This means less cost for business and less hassle for you – everyone is better off. Likewise, if your doctor can obtain more information that helps them to diagnose and treat you better, then that's a good thing too. The downside is that the same data can be used in exactly the opposite way, to remove "undesirables" from the customer base, deny people the best offers, charge them more or exclude them altogether. That's great for organizations wanting to maximize what they do and arguably good for governments trying to maximize overall measures of public good, but not so great for the individuals involved.

Whereas at one time almost anyone could get credit or insurance, these days we live in an increasingly segregated world where some, perhaps even most, get a great deal, but a significant minority get a very poor deal or no deal at all. From a utility maximization perspective that's probably the way it should be, but from a rights-based perspective these things are more questionable – it's back to the chocolate cake again and how to divide things up. Take insurance. Insurance is based on the principle of pooled risk. Everyone contributes premiums, and if someone has a mishap the pool pays out. Imagine what it would mean if an insurer could predict with certainty who will and who won't make a claim. If this were true, then the whole idea of insurance would become pointless for everyone except the insurer. The insurance company will only offer insurance products to those who would never claim, and those that needed insurance would not be able to get it. An extreme example maybe, but this is the direction of travel as more and more is known about us, resulting in better predictions being made about what we are going to do.

An alternative (rights-based/Kantian) view is that data about you is in effect, a part of who and what you are. Therefore it needs to be treated with the same level of respect that you would expect in any normal interaction with another person, business or government body. It being your personal property, you have a right to decide who you share your data with and to what uses it is put. The downside of this approach is that there may be times when people are not managed in the best possible way or are put at a disadvantage because they have not disclosed as much about themselves as other people. Likewise, business suffers in areas like direct marketing because organizations can't assess people as effectively if they don't have as much information.

These two competing views have led to very different legislative approaches to data ownership and usage. In many countries, including the 28 nations of the European Union, with a combined population of more than 500 million people, a rights-based approach to data is taken. EU Data protection legislation is based on a number of principles that in theory place decisions about data use in the hands of the individual. In particular:

- **Consent.** An individual must give permission before an organization can obtain and use data about them.
- **Proportionality.** Data held about an individual should not be excessive, used only for the purposes for which it was gathered and should not be kept for longer than necessary to carry out the intended purpose.
- **Third parties**. An organization can only pass data to a third party if the individual has given their permission for them to do so.
- **Subject access.** Everyone has the right to know what data an organization holds about them.
- **Accuracy**. Data should be accurate and up to date. This does not mean data cannot contain errors, but that reasonable care must be taken to ensure data accuracy and a process must be in place to correct errors when they come to light – for example as a result of a subject access request.

In practice, whenever somebody signs up to an online store, joins a social network, registers with a dating agency and so on, then there will be a statement describing the uses to which their data will be put. What you tend to find is that the data protection statement tends to be worded so that the individual gives broad permissions. Figure 5.1 provides a typical example of a data protection statement used by UK banking institutions.[16]

The main purpose of the data protection statement in Figure 5.1 is to allow the company to manage customers' accounts. The phrases in bold italics (my emphases) give the organization permission to use the information provided by the individual for other purposes. This is important for organizations that offer a wide range of products and services, or who enter into joint ventures with third parties, because it gives permission for information obtained in relation to one product or service to be used to promote other products and services.

Within the bounds of EU data protection legislation, what organizations can't do is things like driving around gathering information about people, for example taking pictures of them in the street or gathering information

---

USE OF PERSONAL INFORMATION

We will store and process personal information about you within the information systems of the company. This includes information we may obtain from third parties such as a credit reference agency, or from other organisations to whom you have previously given permission to share your personal information with us.

We will use this information to manage your account and to provide you with regular information about the status of your account. This includes using your personal information to assess the status of your account or for research purposes, *and to continue to provide you with improved products and services*. We may, from time to time, inform you of new products and services that may be of interest to you.

The information you provide may also be used to make future assessments for credit and to help make decisions on you and members of your household, about any credit or other products or services that we provide, and for debt tracing and to prevent fraud and money laundering.

We may give information about you and how you manage your account to people who provide a service to us or are acting as our agents, on the understanding that they will keep the information confidential.

FIGURE 5.1 A UK data protection statement

from their Wi-Fi signals. Likewise, you can't turn up on someone's doorstep or cold call them, ask questions and record information about them, without gaining their permission first. A prime example of where an organization has upset EU regulators is Google's Street View project. As well as taking pictures, Google also gathered data that could be linked directly to specific individuals. It did this without asking them, and was subsequently ordered to delete it.[17] Concerns have also been voiced by several EU regulators over the way Google (and other online businesses) store and uses personal data in their day-to-day operations.

In some countries a more relaxed view of personal data is taken. In the USA for example, the starting point is very much along the lines of: "Do what you want with data." If there is a specific problem or issue that needs to be addressed concerning storage and usage of data, then we will legislate against it. In practice, this has meant very little data protection legislation in the USA. However, some protections for US citizens are provided under the US constitution. This grants a right to privacy against unwanted intrusion or interference in someone's private life, including unauthorized disclosure of their private information. The key check and balance is that if an individual has suffered harm or inconvenience due to a data issue, then they can sue the organization responsible in the courts.

One area in which specific legislation has been implemented in the USA is around the use of data in credit granting. The Fair Credit Opportunity Act 1970 places legal requirements on credit reference agencies to maintain accurate data, and to disclose what information they hold to individuals that request it. The Act also places restrictions on the types of data that can be used to make lending decisions. For example, marital status, race and age can't be used as reasons to decline someone for a loan. If someone is declined for credit, then the Act obliges lending institution to provide detailed reasons as to why the request was denied. Another area is healthcare. The Health Insurance Portability and Accountability Act 1996 (HIPAA) also grants individuals many similar rights over their data when it comes to medical data.

Another area that requires consideration is the accuracy (veracity) of personal data. If you are using data to decide how you are going to treat people, then you should be reasonably certain that the data is correct so that the right decision is made. Maintaining the accuracy of personal data has always been a big concern for organizations that deal with it. For decades consumer groups around the globe have fought running battles with governments and credit reference agencies over the accuracy of the data that credit reference agencies hold about people. There are countless tales of people being denied credit, refused jobs and otherwise being disadvantaged because a credit reference agency held incorrect data, or incorrectly matched data it held about one individual to another. However, most of these problems have not been due to credit reference agencies acting irresponsibly.

One reason for problems with credit reference data is that the credit reference agencies rely on the organizations that supply them with data to get it right. Another issue is that even if the raw data that the credit reference agencies receive is accurate, there are bound to be some mistakes when you are talking about matching hundreds of millions of pieces of information together, based on name, address and date of birth. Even if a precise match key, such as social security/national insurance number is used, which in theory is unique to an individual, what you find is that people get their numbers wrong or the numbers are used by fraudsters. Therefore there will always be some inaccuracies in the data. Rather than seeking to ensure 100% accuracy, the important thing is to have appropriate mechanisms in place to correct inaccurate data when an organization becomes aware of it.

Maintaining data quality has always been an issue with consumer databases. As a consequence, most organizations have checks and controls in place at the points where data enters their systems. Data is only allowed onto their databases once it has been formatted, cleaned and validated. However, as organizations move into Big Data the accuracy issues become more acute for several reasons:

* The increased volume of data requires significant time to check, process and correct.
* Unstructured data is far more difficult to validate than structured data. With something like date of birth you can check that the day, month and year are all within allowable ranges and that no one has a date of birth in the future or was born hundreds of years ago. But how do you check that a photo or sound sample is valid?
* The data is "raw." Organizations won't let data into their operational systems until it has been appropriately formatted and validated. If you have created the data then you have at least some assurances around the quality of that data. If you are trawling external data sources created by others, then you have no assurance that the data is correct.

The final point is particularly challenging. Let's say, for example, that you find some information about someone on an Internet site. On the basis of this information you decide not to invite them to a job interview or you deny them a product or service because the information suggests that they are a fraudster. If that data is subsequently found to be incorrect, would the individual have a legal claim against you?

5.5   Data security

Hackers steal millions of Sony PlayStation account details.[18] UK government loses data disks containing information about all UK benefit claimants.[19] Payment card processors hacked in $45 million fraud.[20] Council fined £250,000 after employee records found in supermarket car park recycle bin.[21] These are just a few high-profile stories about breaches in data security that have appeared in the press in recent years, and you can bet your bottom dollar that what appears in the media is just the tip of the iceberg. Every year in the UK alone there are thousands of data security breaches recorded by the authorities.[22]

When it comes to data security it's not just about monetary losses from things like stolen credit card details that should concern an organization. Likewise, the fines that the authorities might impose are not necessarily the main issue that one should be concerned about. The biggest area of risk is the reputational damage and loss of customer confidence that can arise when data is released or taken without their permission.

One of the issues with the power of modern technology is that it is so easy to transfer and transport huge amounts of personal data. In the 1980s, I might have been able to copy details of a few thousand customer accounts onto portable media, but today my phone has the capacity to store the account and security details of every credit card in the UK.[23] If I can access a card network online, then my Internet connection allows me to download the same information to my laptop in less than an hour, and an hour after that I've sold it on to criminals overseas. If an organization has open systems, then a single disgruntled employee or one lone hacker can cause a lot of damage very easily.

The risk of data breaches occurring has not necessarily increased, but the potential impact of a data breach has grown considerably. Therefore there is much more incentive to spend more time and resources to ensure that systems are secure. Further complexity is introduced with the move to outsourcing and cloud services. This requires an additional layer of governance to ensure that the service provider adheres to your security policies. This is important because, when it comes to liability, if a cloud provider leaks your data or someone hacks it, it's still you that has ultimate responsibility – not the outsourcer.

5.6 Anonymity

An important issue that has come to the fore in recent years relates to the release of anonymous data. Data protection and privacy laws, such as EU data protection legislation and the HIPAA Act in the USA, only relate to personal data that can be directly associated with a specific individual. It's not personal data if you can't say who the data belongs to. If I have a database containing details of customers' credit card transactions, then that's personal data. However, if I remove identifying features, such as the account number, name and address then it's no longer possible to tell whose credit card details are whose. This is now anonymized data, and as such, data protection and privacy laws no longer apply. Consequently, I can sell or share this data without needing to seek permissions from the individuals in question.

Why do organizations want to share anonymized data? There are many reasons. A government may wish to publish data sets containing sensitive health information so that researchers can seek to find causal factors in people's conditions and the effect of the treatments they receive for those conditions. Credit reference agencies are interested in taxpayer information in order to enhance their credit scoring models (i.e. an individual or company's tax-paying behavior is a secondary behavior when it comes to predicting loan repayment behavior and vice versa). Likewise, marketing companies want to use additional data to refine their customer segmentation models for target marketing purposes.

A key problem is that these days it's no longer enough just to remove obvious personal identifiers such as account numbers, names and addresses. Consider the follow six pieces of information:

- 37 years old
- Female
- Drives a blue BMW 3 series
- Has three children
- Owns a five-bedroom house
- Lives in Boston, Massachusetts.

On their own, each of these is a fairly innocuous thing that has little relevance to any one individual. However, put them together and they form a very specific view of someone, and I would be very surprised if there is more than one individual that has all of these attributes. These days there are so many publically accessible data sources containing all sorts of personal information. By cross referencing what you believe is anonymized data against these data sources it is often possible to identify specific individuals – meaning the data is no longer anonymized.

A number of organizations have fallen foul of this issue in recent years. One well known example is the Netflix prize. In 2006 the movie streaming company Netflix released what it believed was anonymized data as part of a competition with a $1million prize, to see who could build the best model for predicting movie ratings submitted by its customers. However, Netflix was subsequently subject to a lawsuit alleging breach of privacy when it was shown that it was possible to identify some people on the Netflix dataset by cross-referencing it with publicly available information on the Internet Movie Database (IMDb).[24] What was really interesting about this case was that all it took to identify someone on the Netflix dataset was a handful of their movie ratings and the approximate dates when those ratings had been made.[25]

A result of the Netflix case, and others like it, is that organizations are becoming increasingly careful about how they deal with anonymized data.

Once strategy that can be adopted is to slightly perturb the data before it is released. Instead of publishing an individual's actual age as 37 years old, this could be changed to 36 or 38 instead. Likewise, for something like annual income, a random value of up to $100 is added or subtracted from the real value, or values rounded to the nearest $100. In this way the data is close enough to the truth to allow robust data analysis to be performed, but different enough to prevent it being associated with specific individuals. You can also band (group) the data. Instead of saying someone is aged 37 and with an income of $67,950, you describe them as aged 35–40 with an income of $65–70,000. Another approach is to aggregate data. Instead of publishing details about specific individuals, the details of several individuals are combined together – say at the level of a postcode (zip code). One can also remove specific records from a database if it is believed that these are more identifiable than others due to some unusual trait or characteristic.

Methods such as perturbing and aggregating data will reduce risk, but it's probably impossible to guarantee that a data set is truly anonymous. As noted by the Information Commissioner's Office (ICO) in the UK, the risk of anonymized data becoming linked back to specific individuals is essentially unpredictable. This is because one can never fully ascertain what data is already available or what data may be released in the future. It is also infeasible to guarantee the recall or deletion of data (i.e. removing it from a website) once it has been placed in the public domain. You can never be sure it has not been copied to some other database.[26]

One take on this is: it's not worth the risk to share/publish anonymized data sets under any circumstances. However, the approach taken by authorities in the UK is a pragmatic one. If reasonable care has been taken to try to anonymize a data set, then that's sufficient to comply with the law, even if a way of linking the data back to specific individuals is subsequently discovered.

5.7 Decision making

The increasing use of predictive analytics and automated decision making raises a number of questions over how personal data is used in different decision-making processes. Consider the following list of personal characteristics:

- Age
- Type of car you drive
- Smoker (Y/N)

- Units of alcohol drunk last week
- Religion
- Time lived at current address
- Race
- Education
- Gender
- Occupation
- Sexual orientation
- Time in current employment
- Marital status
- Income
- Number of dependents
- DNA details
- Residential status
- revious credit history
- Medical records
- Last book that you bought

Which of these data items do you think it is it ethical to use when making a decision about someone?

How you answer this question depends on the context and the type of decision being made. Let's begin by thinking about this question from a commercial perspective, where data is going to be used by an organization to maximize profit, for example, deciding who to mail with a product offer, what insurance premium to charge, whether to offer someone a job and so on. One perspective is that a person has a right not to have these types of decisions made on the basis of information over which they have little or no control. It is only ethical to make such decisions using factors over which they have had some ability to change and reflect the choices they have made throughout their lives, such as how much they drink, the type of car they drive and so on. Likewise, I shouldn't be denied insurance because of my DNA, I shouldn't be refused credit because of my race and I resent the fact that I get junk mail for stair lifts just because I'm aged over 50! In response to this sentiment, laws exist in many countries preventing immutable data such as age, race and gender from being used in many types of decision-making process.

For some types of personal data, such as religion, sexual orientation and so on, there are questions over whether these are immutable or chosen. People do have some choice in these matters, but there is undoubtedly an inherited component. People are usually of the same religion as their parents, even though in theory there is nothing to stop them from choosing to follow another religion or declaring themselves atheist. Therefore, it is prudent also to question the appropriateness of using this type of personal information in some situations.

If you pursue this line of argument, then you could also question the use of relatively innocuous items such as income or education if there is some evidence of prior prejudice that has led to that situation. If women and men are paid different rates for the same work, then the indiscriminate use of income is questionable. An alternative approach would be to consider male and female income as separate items of data, or to align the two incomes (applying a multiplier) before using them for commercial purposes.

If we now shift the emphasis to something like healthcare, where we want to predict future outcomes in order to decide upon treatment strategies, then it's possible to view the immutability of data from a very different perspective. What you find is that most people do not have a problem with a doctor using information such as age, gender, DNA and so on to make a diagnosis and to decide upon the best possible treatment plan.

So what's the difference between the profit-making organization and the doctor? With the former, decisions are made primarily for the benefit of the decision-maker. Sure, the subject of the decision may also see some benefits, but that's not what it's about. Whether or not the subject suffers harm or inconvenience as a result of the decision is only of secondary concern, if it's a concern at all. Some organizations do put the well-being of their customers at the heart of what they do and operate ethically focused corporate cultures, but very many do not, or pay little more than lip service to their ethical policy if there is any chance of it damaging their bottom line.

There are countless tales of corporations putting profit before people. This includes pharmaceutical companies that bias trials to make their drugs look better,[27] banks systematically mis-selling insurance to people they know will never be entitled to claim,[28] utility companies lying to customers to get them to switch providers,[29] mortgage providers lending to people they know will almost certainly end up in foreclosure and food companies using misleading labeling to sell their products. Doctors, on the other hand, are making a decision to maximize the outcomes for their patients – it's not for their own benefit.[30] The decisions that they make are about more than their own well-being.

The third factor to throw into the pot is the impact of the decision – do the actions resulting from a decision result in a positive or negative outcome for the individual? So perhaps one approach to the ethics of decision making is to consider these three things together, i.e.:

1. **Beneficiary.** For whose benefit is the decision being made?
2. **Impact.** Will the decision have a positive or detrimental effect on the individual?
3. **Immutability.** To what degree can the individual change or alter their data through their lifestyle choices?

This leads to a view of the sensitivity of using data in decision making as illustrated in Figure 5.2.

In relation to Figure 5.2, if someone is taking decisions about people for their own benefit (to make a buck, optimize their business processes etc.), that has a high impact on individuals, and the decision is made on the basis of information over which individuals have no control, then this is the area of greatest concern. It is when making these types of decisions that organizations need to think very carefully about how their actions impact people and where government intervention and legislation are most appropriate. I'm not saying there aren't situations where an organization can use high-impact immutable data to further their own ends, but that they need to tread carefully, and be prepared to respond to challenges on the way that they use that data.

On the other hand, if we are talking about more altruistic decision making, where the beneficiary is the individual, then there is a stronger case for employing all of the data that you have at your disposal in making that decision. Likewise, if the outcome of the decision is relatively benign (such as targeting people for stair lifts on the basis of their age), then there is more of a case for a *laissez faire* approach to how individual data is used.

A somewhat more difficult area is that of social benefit, where governments or other bodies are acting to improve general well-being across the population.

Primary beneficiary	Impact on individual	Immutability of data	Risk of ethically questionable usage
Decision maker	High ↓ Low	High ↓ Low	High ↓
Individual	High ↓ Low	High ↓ Low	↓ Low

FIGURE 5.2 / **Risk of ethically questionable usage**

Some people may suffer harm for the greater good. Classic examples of where there is much disagreement around data usage are law enforcement and national security, and the degree to which governments can passively gather data about law abiding citizens just on the off-chance that the data may be useful for national security purposes at some later date.

Our discussion so far has been about the types of data that should or should not be used to make decisions. A different perspective is that the type of data is irrelevant. What matters is the process by which predictions and decisions are arrived at. If an impartial process is used, which produces an objective measure of how someone will behave, then it is acceptable to use any available data in the decision-making process. It's only when irrational bias/prejudice occurs that the ethics of that decision come into question. If there is some quantitative evidence that, say, men are better credit risks than women or vice versa – then so be it, and a bank's lending policy can be formulated to take this difference into account. This is one of the great strengths of predictive analytics – it is mostly a data-driven process.[31] Data only contributes to a prediction if there is historic evidence that this is the case.

A counter argument against data-driven approaches is that if past decisions were biased (e.g. don't offer jobs to women or minorities) then this will be reflected in the analytics process, and thus the bias is perpetuated. This may be true, but there are ways to correct for bias when a predictive model is constructed.[32] Another factor in favor of data-driven predictions is that there is no intentional bias. Bias may exist, but it's not malicious or unfounded. This can't be said to be true for human decision makers. Even when people intend to act impartially, there are lots of studies that show we all display bias of one sort or another, either consciously or unconsciously. Some of the best examples of this are with job applications and essay marking. If the name of the student is removed from their essay, then students get different marks. What you find is that examiners unconsciously bias their marking towards one gender or another (inferred from the first name), or against people of different ethnic backgrounds (unfamiliar surname). The same effect has also been observed when assessing CVs.[33]

Legislation in many countries is moving away from the view that the method of assessment is more important than the data used to make that assessment. Restrictions are increasingly being placed on immutable data types that someone was born with or can't otherwise change, and some other types of data as well. In 2012, for example, the EU banned discrimination of all types of insurance on the basis of gender, despite strong evidence that when it comes to insurance, gender is a key predictor of the likelihood of making a claim. It is likely that in future years this type of legislation will be extended to cover additional data types.

6

chapter

Types of Predictive Models

Predictive models come in all shapes and sizes. There are dozens, if not hundreds, of different methods that can be used to create a model, and more are being developed all the time. However, there are relatively few *types* of predictive models. The most common ones are:

- Linear models
- Decision trees (also known as Classification and Regression Trees or CART)
- Neural networks
- Support vector machines
- Cluster models
- Expert systems

What you tend to find is that there are quite a few different ways to generate a linear model, several different algorithms for creating decision trees, many ways to derive a neural network and so on. This is a non-technical "formula-free" type of book. Therefore we will not be going into the mathematics of the different methods that can be used to create each type of model. I will, however, say a few words at appropriate points about some of the most popular approaches so that you are not fazed if they come up in conversation at some point.

Before entering into the details of the different models, remember that the methods used to generate predictive models need data. Data is the fuel that drives the analytics process. In particular, two types of data are required within the model development sample:

1. **Predictor data (predictor variables).** This type of data is used to make predictions, i.e. cata that could feature in the predictive model – for example people's income and age.
2. **Behavioral (outcome) data.** This is the behavior that we want to predict.

In order to build a predictive model, the development sample needs to contain both types of data. An appropriate mathematical/statistical technique is then applied to determine what relationships exist between the predictor data and behavior. The relationships that are found are captured in the resulting model. Once a predictive model exists, it can then be applied to new cases where we know what the predictor data is, but where the behavior is unknown (usually because it hasn't happened yet[1]). We can then decide what action to take for a given case, based on the prediction generated by the model.

Although we are not getting into the detail of the techniques that can be used to generate predictive models, it's worth being aware that each method contains its own logic for determining which predictor variables feature in the model and how much they contribute to the final score. The scorecard in Chapter 1 is a type of linear model, constructed using a method called "stepwise logistic regression." However, if another method, such as "Integer programming" or "minimized least squares" had been used, different variables might have been selected as significant and/or different weights allocated to each variable; i.e. each method of model construction generates a somewhat different model, and each model will generate slightly different predictions.

Are these differences important? Probably not in the way you might think. One feature of predictive analytics is the "flat maximum effect."[2] The flat maximum effect states that for most problems there is not a single best model that is substantially better than all the others. Instead, there are potentially millions of models that will give results that are as good, or within a whisker of it, as the best very one.

This is a great feature of predictive analytics. It means that you can experiment with lots of different ways of generating a predictive model and come up with one that you like the look of and satisfies business constraints about what variables should/should not feature in the model, as well as one that generates accurate predictions.

Having said this, for every good model there are many, many bad ones that don't predict very well. So while a degree of *laissez faire* can be adopted, there is still a great deal of expertise required to find a good model amongst all the possibilities.

In the remainder of this chapter we are going to introduce each type of predictive model and discuss the pros and cons of each. We'll also talk about the increasingly popular practice of generating "model ensembles." This is where several predictive models are constructed and the predictions from each individual model are then combined together in some way.

6.1   Linear models

Linear models are the most widely used type of predictive model. The features of a linear model are:

- The relationship between each predictor variable and behavior is represented by a weight.[3]
- The contribution that each predictor variable makes to the model score is calculated by multiplying the predictor variable by the weight.
- The final model score is calculated by adding up the contribution made by each predictor variable (the sum of the predictor variables multiplied by the weights).

Figure 6.1 gives an example of a linear model that predicts a family's weekly household grocery spend at their local supermarket.

Spend ($) =		
19	+	(Model constant)
30 * Income (net monthly $000)	–	(Weight 1 * predictor variable 1)
6 * Distance from store (miles)	+	(Weight 2 * predictor variable 2)
22 * Home owner	–	(Weight 3 * predictor variable 3)
17 * Apartment	+	(Weight 4 * predictor variable 4)
77 * Family size		(Weight 5 * predictor variable 5)

FIGURE 6.1   A linear model

The spend model in Figure 6.1 is a regression model. It predicts *how much* a family spends (as opposed to a classification model that might predict if someone shops in the store next week). A prediction about how much a family spends begins with the model constant of $19. You can think of the constant as the starting point or baseline for the prediction. Points are then added/subtracted from the constant as additional information is added into the model. Income (in $000s) multiplied by 30 is added to the constant, then the distance from the store multiplied by six is subtracted. For a family with a net monthly income of $5,000 that lives two miles from the store, the score at this point is $19 + 150 – 12 = $157.

We then come across some categorical data about the family's residential status (home owner). However, this isn't numerical data. The statement "22 × Home owner" is nonsensical! The way that this type of data is dealt with is for an "indicator variable" to be created. The indicator variable takes a value of one if the individual is a home owner and zero otherwise (i.e. if they rent or live with their parents). Twenty-two points are added to the score if the person is a home owner (22 × 1) and 0 otherwise (22 × 0).

The same principle also applies to the next predictor variable "Apartment," which indicates if someone lives in an apartment, rather than another form of housing. An indicator variable for apartment has a value of one if the family lives in an apartment and zero otherwise, and this value is then multiplied by –17 and added to the score. Finally, family size multiplied by $77 is added to get the final model score. If the family are home owners, they don't live in an apartment and there are four of them, then the predicted weekly grocery spend using this model is $487.

So far so good, but as it stands, the model in Figure 6.1 has a number of issues. One problem is that for a linear model, if it's going to predict well, the underlying relationships between each predictor variable and the outcome variable must be linear (or at least approximately so). Linear simply means that if you plot the relationships between each of the predictor variables and the behavior being predicted, you get something that's not a million miles from a straight line. To illustrate this, let's examine the relationships between grocery spend and two of the predictor variables, as shown in Figure 6.2.

In Figure 6.2, the relationship between Family Size and Spend more or less follows a straight line, but the relationship between Income and Spend does not (as shown by the curved trend line). In technical speak we say that Family Size has a linear relationship with Spend and Income has a non-linear relationship with Spend.

One of the big criticisms of linear models is that they will only create accurate predictions if the relationships in the data are linear. This is completely true.

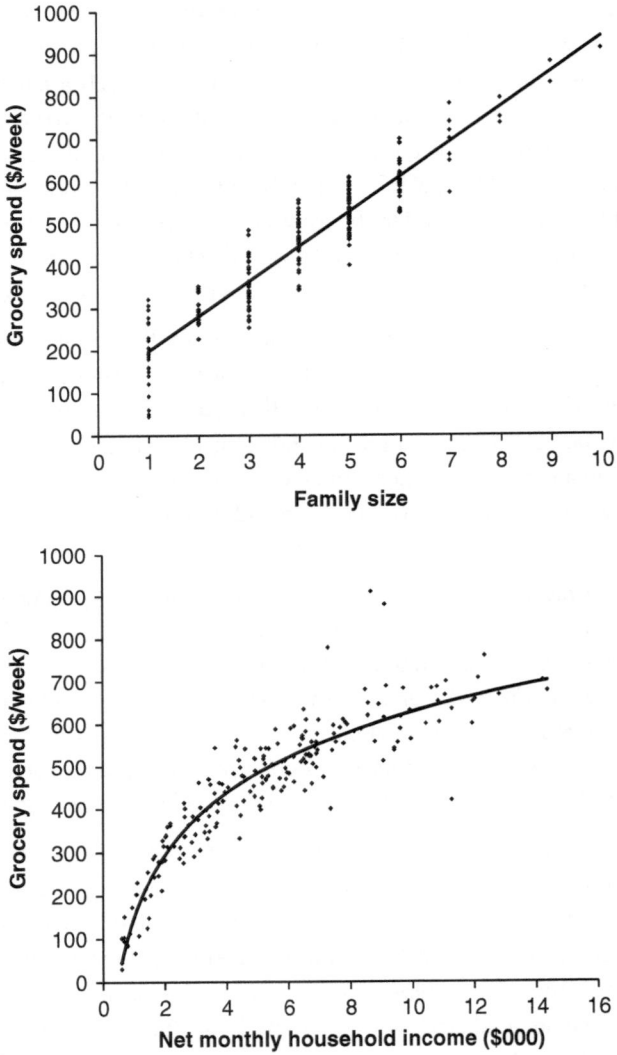

FIGURE 6.2 / Linear and non-linear relationships

However, there are a number of things you can do to overcome this problem, and it is a serious (but common) mistake to dismiss linear models simply because you have identified some non-linearites in the raw data. When faced with non-linear data the way to deal with it is to transform it into something else that is linear. Maybe Income doesn't display a linear relationship with spend, but the square root of Income does (this is true for the Income data in Figure 6.2). Therefore you use the square root of Income in the model instead.

Data transforms, such as replacing the raw variable with the square, square root or log of the variable, are a popular approach, but the most practical way of dealing with non-linear data is to discretize (bin) the data. Instead of including raw Income, Income is divided into a number of ranges (bins). A 1/0 indicator variable is then used to represent each range. If the person has an income within a given range the indicator variable is set to 1, otherwise it is set to zero. This s the same approach that we took to categorical data, representing each category with a separate indicator variable.

For the model in Figure 6.1, let's assume that it makes sense to bin Income into five income ranges[4] ($0–2,400, $2,401–4,200, $4,201–5,900, $5,901–9,300 and $>9,300). Five indicator variables are created, one to represent each range.[5] The indicator variables are then used to construct the model instead of the raw value of Income.

As well as dealing with non-linearity binning removes another big problem: missing data. You may not know the income for some people, where they live, or how many people are in their family. In the original model in Figure 6.1, you would not be able to calculate a score if the Income, Distance, or any of the other data were missing (and if it's missing you can't assume its zero!) With the binning approach one simply creates another 0/1 indicator variable to represent missing data.

A very common and successful approach that is adopted when developing predictive models is to discretize all the data. Every numeric variable is segmented into ranges, and each range is replaced with a 0/1 indicator variable. So for the model in Figure 6.1, even though there is a linear relationship between Family Size and Spend, Family Size would be cut up into a number of ranges, and indicator variables created for each range. If the model is now constructed using the indicator variables, then what you end up with is a scorecard-type linear model – as shown in Figure 6.3.

Binning all the predictor variables and replacing them with a set of indicator variables has several other advantages. One is that it standardizes the data. This makes it much easier to compare the contribution that each variable makes to the model score. In Figure 6.1 you can't say whether having a net monthly income of $8,000 contributes more or less to the model score than being a homeowner without doing some calculation. However, with the scorecard version you can see immediately that if your income is $8,000 it contributes +172 points to the score, compared to +23 points if you are a home owner.

A further advantage is that all you have to do to calculate the score is add up the relevant points, rather than doing some multiplication first. To someone

Constant	$19
Income (net monthly $)	
0–2,400	88
2,401–4,200	121
4,201–5,900	147
5,901–9,300	172
9,301 or more	190
Distance (miles)	
0–2 miles	–5
3–5 miles	–24
>5 miles	–36
Home owner	
Yes	23
No	0
Appartment	
Yes	–16
No	0
Family size	
1	68
2–3	193
4–5	303
6+	539

FIGURE 6.3 A linear model using indicator variables (a scorecard)

with a technical background this sort of thing may seem pointless. But to someone who doesn't have that background, the scorecard version of the model is conceptually much simpler to understand than the model in Figure 6.1.

Another issue that often comes up when building predictive models is outliers. Most families have annual incomes in the range $20,000 to $250,000 but a (very) few have incomes of $10m or more. It doesn't really matter if these high incomes are real or a mistake – what tends to happen is that they distort the results. In traditional statistics a popular solution is to simply exclude these types of cases and build your model without them.[6] However, in predictive modeling it's often the outliers that are of most interest, particularly when models are being constructed to predict "rare" events such as fraud. Binning automatically takes account of outliers, and treats them appropriately within the model.

Despite its many advantages, binning is not popular with everyone. One challenge against binning is that some granularity in the data is lost. For the model in Figure 6.3, you get the same score if your monthly income is $6,000 or $8,000, but the version in Figure 6.1 would give different scores for these different amounts. This is a valid point, but in all of the experimentation that I have undertaken in the past (and most of my colleagues and peers agree), as long as you have a reasonable number of well-defined ranges, then the binning approach does not result in any deterioration in model performance. If anything, the loss in granularity is more than offset by the benefits. In my experience, the use of binning always results in better linear models than when the raw data is used or when simple transforms such as log or power functions are applied.

For classification models, where the score represents the probability that someone will do something or not, the most popular method for generating linear models is a process called Stepwise Logistic Regression.[7] "Stepwise" means that a model is constructed in stages. Initially, a model is constructed using just one variable that the algorithm determines is the most significant predictor of behavior. At each subsequent stage additional predictor variables are added (or removed), based on how much or how little the additional variables add to the predictive ability of the model – given the variables already in the model. The process stops when no more predictor variables are found that improve the model. Consequently, only predictor variables for which there is statistical evidence that they contribute to the predictive ability of the model are selected.[8]

A great feature of the stepwise process is that if two variables are highly correlated, only one is likely to feature in the final model. This is because the other is unlikely to add significantly to the model, given that the first variable has already been included.[9] Another great feature of stepwise methods is that you can see the order in which variables entered the model, i.e. which variables add most and which only make a marginal contribution.[10]

As well as Logistic Regression, other popular methods for generating linear (classification) models include Discriminant Analysis (the first popular method for developing Predictive models), Linear Regression (least squares),[11] Integer Programming (a method championed by FICO for building credit scoring models), Genetic Algorithms, probit and Tobit analysis (Both similar to logistic regression) and Bayesian methods.

For regression problems, where one is trying to predict the magnitude of behavior (how much, how long etc.) the most popular method for generating linear models is Stepwise Linear Regression. Linear regression has many similarities with logistic regression, but uses a different procedure for determining the model scores. Logistic regression (despite its name) cannot be used for creating regression models.[12] Other popular methods include Least Angle Regression (LARS) and Genetic Algorithms.

6.2   Decision trees (classification and regression trees)

Decision trees are probably the second most popular type of predictive model in use today. The key features of decision trees are:

- The model is created by segmenting a population into smaller and smaller segments.
- The model can be represented as a "Tree diagram."
- The model score is determined by the end node into which an observation falls after passing through the tree.

An example of a decision tree was introduced in Chapter 2, as part of the Booles case study, where it was used to predict response to marketing activity (A classification tree). Figure 6.4 shows another example of a decision tree, this time to predict weekly household grocery spend (a regression tree).

In Figure 6.4 the properties of the decision tree are illustrated using a sample of 100,000 families whose weekly spend is already known. The node at the top of the tree shows the average family spend for all 100,000 cases, before any segmentation has been applied, i.e. $411. The tree "grows" from the top down. Each split in the tree is defined using predictor variables that the decision tree algorithm has determined best discriminate between different values of spend.

For this tree, the segmentation algorithm has decided that the best option is to begin by splitting the population into five groups based on family size – and you can see from the diagram the number of families in each group and the average spend of those families. The splitting process is then applied again to each of the resulting segments. For single people (family size one) the population is segmented into two subsidiary segments based on their income. The algorithm then decides that no further splits are required. So what we are saying is that for people who live on their own, and whose monthly net income is less than $3,750, then the evidence is that these people spend an average of $98 a week on groceries. Likewise, for single people with an income greater than $3,750 the average spend on groceries is $229.

Decisions about how to treat someone are based on the properties of the end node into which they fall (the end nodes are highlighted and numbered in

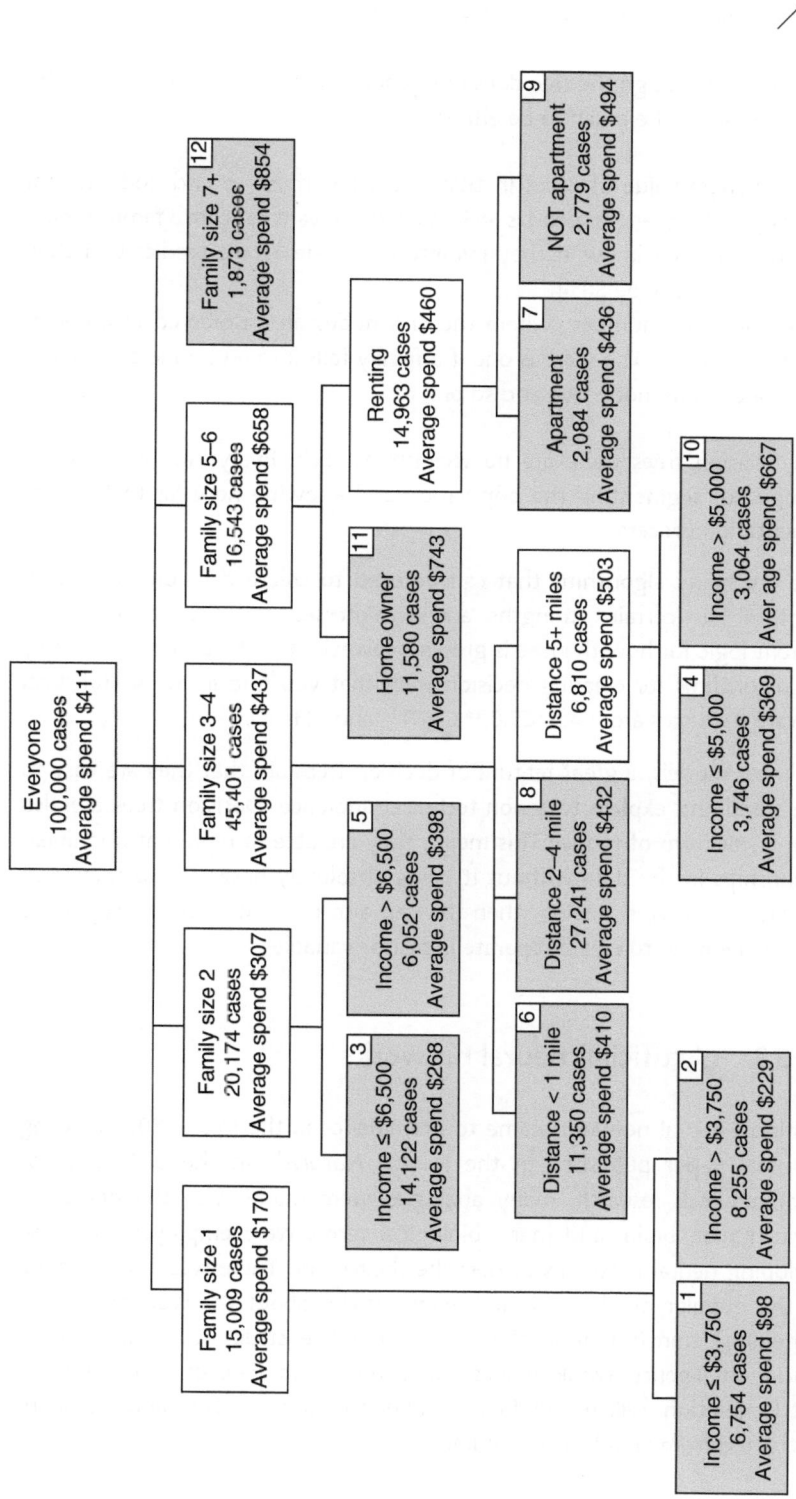

FIGURE 6.4 A decision tree for grocery spend

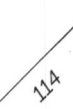

Figure 6.4). Although the tree does not generate a score as such, the score for an individual can be taken to be either:

- The average value of cases in that node. For those in end node 12, for example, the average spend is $854. When we want to score a family whose spend we don't know, if they fall into end node 12 we predict that their expenditure will be $854.
- The end node number (where the end nodes are numbered in order of average spend). The score is one if a family falls into end node one, two if they are in end node two and so on.

With decision trees there are no weights as such. Everything is based on the logic for segmenting the population and knowing how far to take the segmentation process.

There are many algorithms that can be used to create decision trees. Each algorithm has certain strengths and weaknesses and uses a somewhat different logic for how the tree is grown. However, the three most commonly used algorithms for creating decision trees that you find in many statistical software packages are: C4.5/C5.0,[13] CART[14] and CHAID.[15]

Like linear models, a great feature of decision trees are that they are easy to understand and explain to a non-technical audience. Decision trees are also a non-linear form of model. This means they are able to represent non-linear relationships in the data without it being absolutely necessary to transform the data.[16] If data is missing, then this can also be used to define segments without the need to create separate indicator variables.

6.3   (Artificial) neural networks

(Artificial) neural networks came to prominence in the mid-1980s following a seminal paper published in the journal *Nature*.[17] In the early days of neural network research, many analogies were made with the operation of the human brain, and many biological terms were employed by those developing neural networks to describe their work. There was a lot of hype and the popular media published many articles about the wonderful (and frightening) transformation that these brain-like structures, imbued with artificial intelligence, would unleash upon the world. Fans of "Star Trek: The Next Generation" will no doubt remember the android Data and his brain, constructed from an advanced neural net.

It's exciting to talk about neural networks in this way, but in reality neural networks do not possess intelligence and it's wrong to think of them as artificial brains. Like most things, once you know how they work, the mysticism surrounding neural networks quickly disappears.

The key component of a neural network is the "neuron." The operation of a neuron is illustrated in Figure 6.5 – again using the example of weekly family grocery spend.

The operation of a neuron is a two-stage process. In the first stage each predictor variable is multiplied by a weight and the values added together to generate a score. To all intents and purposes this is a linear model, no different from the linear models discussed earlier in the chapter. In the second stage the score is transformed using an "activation function" so that it lies within a fixed range.[18] This is often 0, +1 or –1, +1. The transformed score is the output produced by the neuron.

To create a neural network model, several neurons are connected together. The outputs from one layer form the inputs to the next layer. Figure 6.6 shows a neural network model that predicts weekly family grocery spend.

The network model in Figure 6.6 provides an example of the most common type of neural network used commercially. It contains an input layer, a hidden

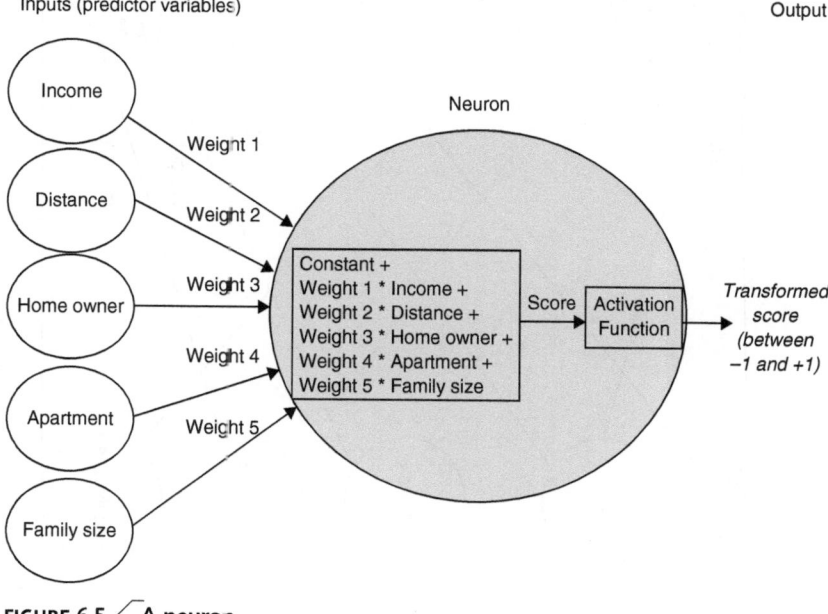

FIGURE 6.5 A neuron

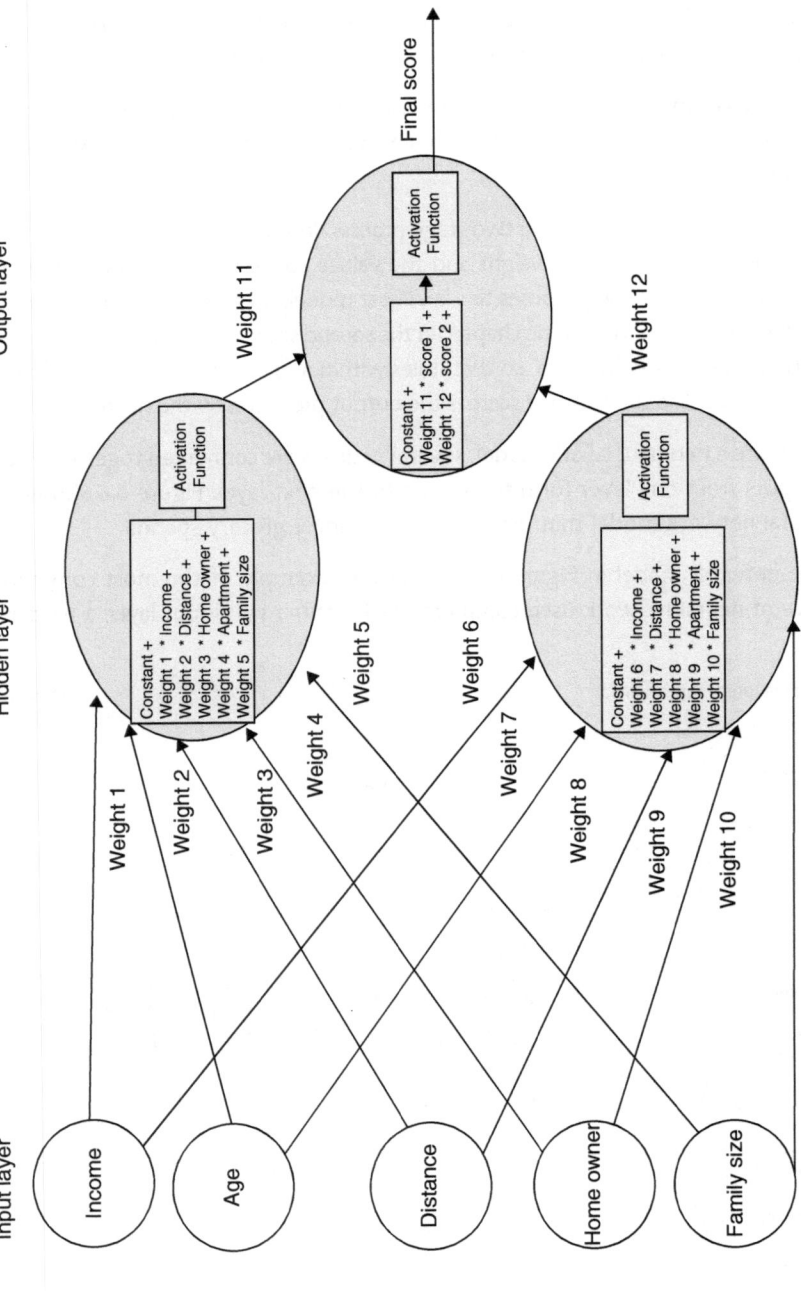

FIGURE 6.6 A neural network

layer and an output layer. The input layer does not contain any neurons as such. It just represents the predictor variables that are presented to the network, with one input for each predictor variable. The number of neurons in the hidden layer depends on the problem. In this example, to keep things simple, there are just two neurons in the hidden layer. In practice there are often between two and twice the number of input variables (so in this example somewhere between two and ten). The output layer contains a single neuron.

The final model score is produced by the single neuron in the output layer. The output layer neuron works in an identical manner to the two hidden layer neurons. The only difference this time is that the inputs to the neuron are the two scores from the hidden layer neurons, instead of the original input variables. The output layer neuron combines the two scores using a linear model. If we had three neurons in the hidden layer, then there would be three scores to combine, four neurons four scores and so on.

The final task of the output layer neuron is to transform the score via its activation function. How this occurs depends on whether the network is designed to predict the probability of an event (a classification problem) or a quantity (a regression problem). For classification the activation function in the output layer neuron would be chosen so as to force the score to lie in the range 0–1, representing the probability of the event occurring. The network in Figure 6.6 is dealing with regression, and therefore a different activation function is applied so that the score is a direct estimate of weekly grocery spend.[19]

An easy way to think about the operation of the network in Figure 6.6 is as a function of three separate linear models. The two neurons in the hidden layer each produce a linear model. The score from each of these is then transformed so that it lies within a fixed range, and these transformed scores are the predictor variables that feed the neuron in the output layer.

The big strength of neural networks is their ability to take into account non-linear features in data, and if you build a network correctly it has the potential to outperform linear models and decision trees in some situations. Their main drawback, as you may have gathered, is their complexity. Figure 6.6 contains just five input variables, two hidden neurons and one output neuron, and that's complex enough. Imagine what a network with 200 input variables and a few dozen hidden layer neurons would look like!

The hard part with neural networks is determining how many neurons to have in the hidden layer and what the weights should be, and there is usually a degree of trial and error involved. You can't determine the weights in a network using a simple formula. Instead, algorithms are applied which run through the data many times, each time adjusting the weights based on how well the model predicts. As more runs are undertaken, so the predictions get better and the weight adjustments become smaller, until no further performance improvements are seen. The original neural network algorithm presented in the aforementioned *Nature* paper was called Back-Propagation. This method is still employed on occasion and works well, but has generally been superseded by more efficient algorithms that are many times faster, and which don't require as much data scientist input to tweak the algorithm's parameters.[20]

6.4   Support vector machines (SVMs)

Of all the models types described in this chapter, support vector machines (SVMs) are arguably the most enigmatic. They share many similarities with neural networks, but are quite hard to explain without resorting to the requisite mathematics.

When talking about classifying people's behavior, such as responding to a marketing communication (event) or not (non-event), one way to think about it is as if the events and non-events are different items that have been scattered across a table. Someone has stood at one end of the table and tossed the events onto that end, and someone else has thrown the non-events onto the table from the other end. Most events/non-events remain towards the end where they were scattered from, but there is some overlap in the middle and one or two might even reach the other end. Your job is to draw a line that results in as many events as possible lying on one side of the line, and as many non-events as possible lying on the other. Two examples of this are illustrated in Figure 6.7.

Figure 6.7(a) shows the best straight line that separates the two groups. If you measure the distance between each point and the line, then sum up these distances, the sum is the most it can be. The line is said to: "maximize the margin." You could draw the line in other places, but the sum of the distances would be less – the margin would not be maximized. In practice, this would mean getting more cases on the wrong side of the line. Conceptually, linear models are analogous to the straight line in Figure 6.7(a). In Figure 6.7(b) the same job is done by the squiggly line, but because the line is not restricted to being straight, it does a better job of separating the two groups. A non-linear method, such as a neural network, is conceptually more like this example, and is why in theory neural networks and support vector machines are superior to linear models.

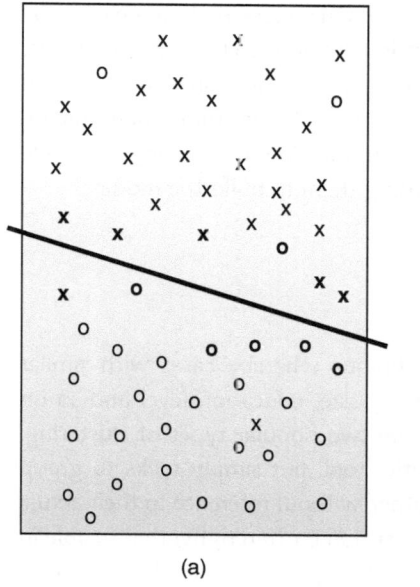 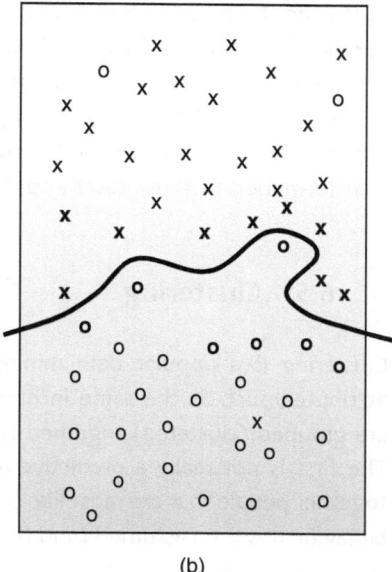

(a)   (b)

FIGURE 6.7 Maximizing the margin

Like neural networks, support vector machines are a non-linear method – in our lay terminology they also use a squiggly line to separate the two types of behavior. However, support vector machines approach the problem in a slightly different way from neural network algorithms. The support vector machine works to find the equation of the line that maximizes the margin, and the equation of this line is your model. The score is then a measure of how likely a case is to lie on either side of the line (high score = probably an event, low score probably not an event). It sounds simple enough, but with situations like Figure 6.7(b) transformations have to be applied before the equation of the line that maximizes the margin can be found. In practice, these transformation are complex, and the structure of a support vector machine model is not interpretable in any meaningful way: you can't describe the results in terms of this variable contributes so many points, that variable so many points, and so on.

When building most types of predictive model, including linear models, decision trees and neural networks, all of the data in the development sample is used to build the model. With a support vector machine only the cases that are closest to the line are used. These are the "support vectors." If you think about it this makes sense. Cases far from the margin (near the edge of the table where the events were scattered from) are not going to tell you much about where the line should be drawn. Only cases that are near to the middle of the table will have a significant impact on the classification process and are therefore included in the modeling process (these are the circles and crosses in bold in Figure 6.7). A key part of the SVM algorithm is determining which cases in the sample are the support vectors, before applying a suitable transformation to those cases and then using them to build the model.

6.5   Clustering

Clustering is a popular data mining technique whereby cases with similar attributes, such as the same income, family size, education level and so on are grouped (clustered) together. There are two popular types of clustering. The first is not really a predictive analytics tool, but simply seeks to group together people that are most like each other, without reference to their actual behavior. It is then assumed (and the assumption is often right) that people in the same cluster will buy the same types of things, have similar health issues, go to the same sort of places on vacation and so on. Consequently, they can all be treated in the same way. This type of clustering is very widely used in marketing, but also has applications in many other areas. An example of clustering is shown in Figure 6.8.

Figure 6.8 shows how the customers of a company selling golfing products are distributed by their income and age. You can see that there are three natural groupings in the data, young people with low incomes, young people with high incomes and older people with high incomes. Golf is quite an aspirational game, so it's not unreasonable to conclude that with the younger people there is one group which contains genuine high flyers who have already carved out a career for themselves, and another group who don't have as much money, but want the association with high flyers that they believe playing golf imbues them with. The company then uses the clusters to decide whether to market customers with offers for premium or budget brands of golf products. What the graph also shows us is that by the time golfers get older, the lower income group has all but disappeared. They have either progressed with their careers, stopped playing golf or taken their business elsewhere.

For this type of unguided (without behavior) type clustering, the most popular clustering algorithms are K-means clustering and Hierarchical clustering.[21] The great strength of this form of clustering is that you don't need to know anything about behavior to do it. You just need a sample of cases and some data items to drive the clustering process.

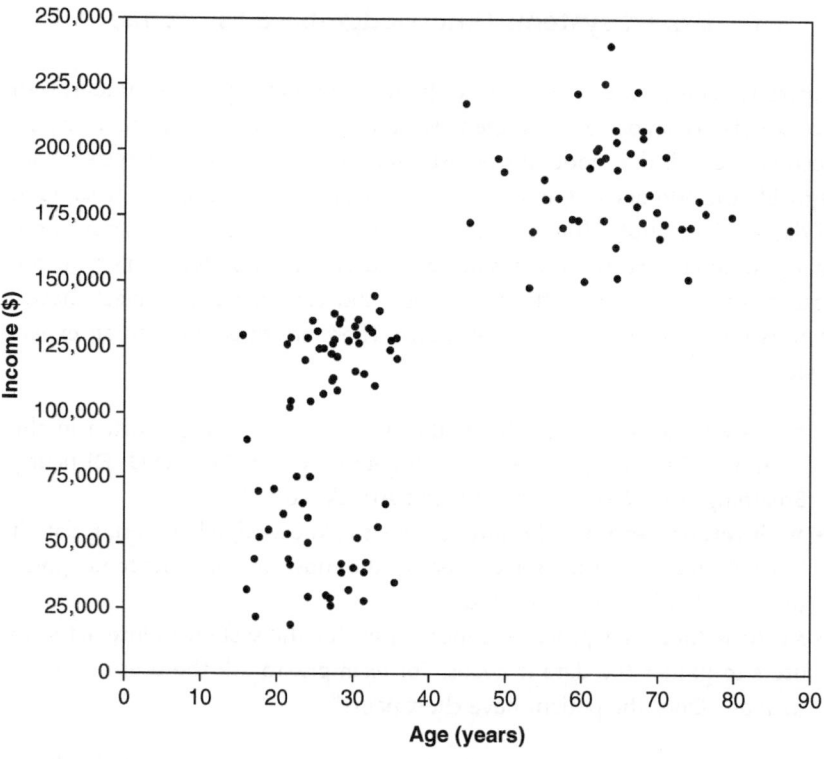

FIGURE 6.8 Clusters

For predictive modeling a slightly different clustering process, called *K*-nearest neighbor, is widely applied, and for this type of clustering you do need to know something about behavior for cases in the development sample. *K*-nearest neighbor is mainly used for classification, but adaptations do exist that can also be applied to regression. At the start of the algorithm the model developer specifies the value of *K*. When a prediction is required for a new case the algorithm finds the *K* cases in the development sample that are "most like" the case for which a prediction is required. The score for the new case is calculated to be the proportion of the *K* nearest cases in the development sample that displayed the behavior. So if *K* = 50, then the algorithm would find the 50 cases most similar to the case you want the prediction for. If ten out of the 50 cases display the behavior, the score is 0.2 (10/50). If 40 cases display the behavior then the score is 0.8 (40/50), and so on. A key question with this type of clustering is: what value of *K* to use? With small samples it is usual to use quite small values of *K* and possibly a value of just one (so you get a score of just zero or one).

6.6   Expert systems (knowledge-based systems)

Expert systems attempt to replicate the decision-making process undertaken by experts. A layperson can then use the expert system to make expert decisions. Of all the model types discussed in this book, expert systems are arguably the odd one out, in that they are not reliant on having large samples of digitized, computer friendly, data in order to build them. That's not to say that you don't need to obtain data to build an expert system, only that the data comes directly from the knowledge and experience of subject matter experts rather than from a digital source. There are three parts to an expert system:

- **A knowledge base**. This holds the decision making logic, often in the form of IF/THEN type rules. For example: IF "Chest Pain" AND "Difficulty Breathing" THEN Probability of Heart Attack = 0.17.
- **An inference engine**. This interrogates the knowledge base, using data it has collected from the user to identify the most probable outcome, given the data that has been provided.
- **An Interface**. This provides a mechanism for the system to interact with users to gather the data it needs. For example, to ask the user questions such as: "Does the patient have chest pain?"

An expert system can contain just a few dozen rules in its knowledge base, but a complex expert system, such as one used to support a GP in diagnosing medical conditions, will contain thousands of rules. Sometimes the rules are fairly simple like the one mentioned previously, but the rules can be very complex and/or interconnected. One role of the inference engine is to move through the rules, asking relevant questions along the way to guide the process. So the first questions it may ask is "Does the patient have Chest Pain?" If the answer is yes, then it goes on to ask about breathing difficulties and so on. If not, then a different line of questioning is pursued. The other role of the inference engine is to weigh up the evidence that has been collected from processing the rules, deal with inconsistencies and conflicts between different rules, and then combine it all together to come to a conclusion that is presented back to the user.

A key role of the inference engine is to ask the right questions in the best order possible, deciding which question to ask next based on the information that has already been gathered. With a heart attack, one would expect to need only to obtain a handful of answers in order to make a diagnosis. The system should not need to run through hundreds of questions to get there if the right questions are asked. So if the patient has already indicated chest pain and breathing difficulty, it would make sense to focus on other relevant factors, such as age, gender and previous medical history (Young and asthmatic would tend to indicate asthma attack rather than a heart attack, while old, male and overweight would lend further support to the problem being a heart attack.)

The hardest part when developing an expert system is creating the knowledge base. To do this, the system developer talks to experts and captures their decision-making logic in a form that can be represented by the knowledge base. This may take many months or years for something complex like a medical diagnosis system that has many thousands of rules, and this is one of the main weaknesses of expert systems. They need a lot of time and effort to create, and the knowledge base needs a lot of maintenance in order to remain current. They are also only as good as the experts(s) from which knowledge has been taken. As well as medical diagnosis, other applications of expert systems include mortgage risk assessment, geological analysis to identify mineral deposits and fault detection in complex machinery.

Expert systems are appealing because the reasoning behind their decision making can be explained easily using the rules that have been applied, and follows human decision-making principles. In the 1980s there was a lot of enthusiasm for expert systems, but they tend to be one of the least used methods in predictive analytics. This is mainly due to their cost to develop and maintain, but also due to the fact that they try to match the expertise of human experts, whereas other predictive modeling approaches aim to do better.

Expert systems may only have a few applications, but where there is very little hard data available to build a model using a statistical/mathematical approach, the idea of capturing human expertise still has merit. With the Delphi Method, a group of experts is brought together and their expertise is used to construct a simple, usually linear, model. The model won't have the sophistication of a full expert system, but it's possible to develop a model very quickly, usually in just a day or two.

There are two stages to the Delphi process. In the first stage, a facilitator gets the group to come up with the ten to 20 or so predictor variables that they think are the most important for predicting behavior. In the second stage the group assigns weights to each variable by consensus. The result is a linear model that looks just like one developed using a method such as linear or logistic regression.[22] The only difference is that with the Delphi method the experts have decided the weights, rather than a computerized process being applied. What you tend to find is that the performance of Delphi models isn't as good as statistically derived models, but they are not bad, and certainly better than no model at all.

6.7   What type of model is best?

When assessing how good a predictive model is there are several criteria to consider:

* Predictive accuracy
* Explicability
* Simplicity
* Stability
* Business acceptability

Predictive accuracy is what most people (rightly or wrongly) tend to focus on most. There have been lots of studies that have looked at the predictive accuracy of different types of models and the different algorithms used to construct them. The conclusions that can be drawn from these studies are:

* On balance, support vector machines and neural networks are slightly better than other types of model for predicting consumer behavior, but often there is not much in it. Most algorithms perform similarly well in many situations[23] (the flat maximum effect again).
* No single type of model is always best. Sometimes a simple linear model or decision tree will outperform more advanced/complex methods.[24]

The difference in the predictive accuracy of different models is usually pretty small.[24] This is true even for problems that are described as being highly non-linear or there are a lot of interactions between variables[25] *as long* as suitable data transformations have been applied (e.g. binning and the use of indicator variables[26]). A classic case is fraud detection. A very widely expressed belief is that you have to use a neural network or support vector machine if you want to produce a decent model, because of the complexities of the relationships in fraud data. This is a misconception, based on the fact that one of the earliest fraud detection systems just happened to be based on a neural network model. I have come across more than one example of industry-leading fraud detection systems based on linear models and/or rule sets that have performed as well as or better than competitors based on more advanced methods. Having said this, one should be careful not to confuse general and specific findings.[27] There is a lot of evidence that a wide range of algorithms yield very similar levels of performance on average, but for some specific problems one method may be substantially better than another – but you can't tell if this is the case until you've built the model. Therefore it often makes sense to develop a number of competing models using different methodologies in order to see which one generates the best model for your particular problem.

One drawback of neural networks is that it is notoriously easy to over-fit to the data, making them appear to perform much better than they really are, i.e. their performance in real-world usage is inferior to their performance based on the data used to develop them. They also require a lot more computer power to generate than linear models constructed using linear or logistic regression, or decision trees using C4.5 or CHAID (often 10–100 times more), which can cause problems when one is dealing with large samples and lots of predictor variables.

Decision trees, like neural networks, are prone to over-fitting and have some other drawbacks. In particular:

- Popular algorithms for deriving decision trees are not very efficient at utilizing data. Consequently, their performance is sometimes (although not always) marginally worse than other types of predictive model for a development sample of a given size.[28] This is particularly true when small and medium-sized samples are used to construct the model.[29]
- For classification you need equal numbers of cases that do/do not display the behavior to build good decision trees. If you have lots more examples of behavior or non-behavior in the development sample then model performance will be poor (e.g. the results of a mailing campaign where only 1% of those targeted respond). The greater the degree of imbalance the worse the model will be. Decision tree algorithms are more sensitive to imbalance than almost any other type model construction method.[30] There are however, ways of getting around this problem.[31]
- The range of scores is smaller than many other types of model, resulting in score distributions that are "clumpy." This isn't necessarily a problem, but can put some people off using them.[32]

For many (and possibly most) applications of predictive analytics, models must be simple and explicable. One needs to be able to say that someone obtained a particular score because they had characteristics A, B and C. In the US, The Equal Credit Opportunity Act 1974 gives everyone the right to know why they were declined for credit in terms of which predictor variables resulted in them getting a low credit score.[33] If you develop a model to predict who to lay off in a time of recession, then you need to be able to explain the decisions if you want to avoid the wrath of the unions and possible legal challenges. Decision trees, linear models and expert systems are easy to explain, but a significant weakness of clustering, neural networks and support vector machines is their complexity and "black box" nature. You can't tell by looking at these types of model what variables contributed significantly to the model score and which did not. It's quite possible that some variables contribute nothing at all, despite being included in the model. There are methods that can be used to infer what variables are important in a neural network, but that arguably just adds another layer of complexity.

The main issue with clustering is that there is no model as such. What comes out of the process is the original data set that you used to do the clustering, plus the cluster into which the observation has been placed.[34] When you have a new observation that you want to assign to a cluster you have to run it through the algorithm again in order to allocate it to a cluster. What you can't say with any certainty is that a case is in a given cluster because of feature X, Y or Z. Instead, one is usually limited to discussing the average properties of the cluster into which someone is placed. In cluster one the average age of people is 30 and most people rent; in cluster two the average age is 50 and most people own apartments. However, there will be a distribution of ages in each cluster, and just because you are 50 and live in an apartment, doesn't mean that you will automatically end up in the second cluster – you are just more likely to.

Most clustering algorithms also struggle with large quantities of data, both in terms of the number of records and the number of predictor variables. However, some of the newer types of clustering algorithms get around this problem by breaking the problem up into parts and/or creating crude approximate clusters as an initial stage, which are then refined as a secondary process.

With regards to model stability, the performance of all predictive models tends to deteriorate over time. Some models last for years before they lose their ability to predict; others have lifespans measured in hours. For operational models that are intended to be in place for months or years before being replaced[35] an often neglected issue when deciding what type of model to use is the expected lifespan of the model. If one type of model needs to be rebuilt every three months, yet another lasts two years before being replaced, then the additional redevelopment costs need to be taken into account when comparing the business benefit of the two alternatives.

So even if the first model is more predictive when it's first built, it may not be the most cost-effective if you have to spend more time and effort maintaining it. There hasn't been a huge amount of study into the stability of different types of models, but what little evidence there is tends to support the case that simple linear models and decision trees are more stable than other types of model.[36]

A final consideration is: Will the business accept the model? One part of this question relates to the physical infrastructure required. Building a model is one thing; integrating it into the business is another. The second part of the question is about people and culture. For many applications of predictive analytics there are those, other than the data scientist who built the model, who have an interest in what variables are important and how a model generates its predictions. As discussed in Chapter 3, it's quite common for people in authority to hold views such as: "No model should ever contain more than 20 variables" or that: "The model can't be any good because gender doesn't feature" or: "The model's a joke! How can anyone believe that these variables predict ..."

Perhaps the biggest mistake that can be made by a data scientist is to think that predictive accuracy is always the most important consideration. Sometimes it is, but for most business applications other concerns take precedence. Consequently, linear models and decision trees still dominate the world of predictive analytics. Their (sometimes) marginally worse performance compared to neural networks, support vector machines etc. is outweighed by their simplicity, stability and the ease with which they can be understood and explained to non-technical people.

Neural networks, support vector machines and other complex methods tend come into their own when there is little need to explain the model to management or the individuals being assessed by the model.

When considering what type of predictive model to use, my advice is:

- Always use linear models as your benchmark, developed using stepwise linear or logistic regression, and replace the raw predictor variables with indicator variables.[37] With the right software, these are quick to develop and will provide a baseline against which to assess other types of model.
- If you can, do a little experimentation. Build different models and compare the results using an independent sample that was not used to construct the model, and ideally taken from a different point in time than the model development sample (an out of time sample).

• Even if you don't want to use a "black box" method such as a support vector machine or neural network, it's often worthwhile developing such a model just to get a feel for how much additional benefit you might get, i.e. how much you are missing out by not using these methods.

In addition to these considerations, you should always consult with the model user as to their expectations: what is acceptable in terms of the nature and explicability of the model and what is out of bounds. We will discuss these issues in more detail in the next chapter.

6.8   Ensemble (fusion or combination) systems

If you get a bunch of experts in a room and ask them to make decisions, you will find that on average that the group decisions tend to be better than any single expert would make on their own. Even if you take someone who is considered to be the very best in their field, and compare their performance against a mixed ability group of their peers, then the consensus view usually wins out. This is because even the best experts don't know everything and are subject to human biases. They can be swayed by what happened in one or two odd/extreme cases, which is not borne out by the wider experience of their peers. If one or two of the experts have a mistaken view, or their judgment is flawed in certain cases, then this will be corrected by the majority.

The same thing happens with predictive models. Every model has its strengths and weaknesses – is very good at predicting certain types of cases but not others. If you create lots of models in lots of different ways and then combine them together, you often find that the ensemble of models is more predictive than any of the individual contributors. This seems to work even if there is a spread in terms of the quality of the models that make up the ensemble.

There are many ways of creating an ensemble of predictive models, but there are three general approaches:

• **Parallel ensembles.** Several models are constructed, each in isolation from the others. The scores from each model are then combined together to generate a final score.
• **Multi-stage ensembles.** Models are constructed in stages. At each stage, the way the model is constructed is dependent upon the results from the previous stage(s). Therefore, the models are dependent upon one another.

As with parallel systems, the scores from all the models are combined together to make a final decision.
- **Segmentation ensembles.** The population is segmented and a separate model constructed for each segment. Unlike the other two approaches, the final decision is made using a single model, depending upon the segment into which an individual falls. For the experts in a room, this would be like having a triage process to assess each case to decide which expert would be best at dealing with it.

Where there is a requirement for models that are simple and explicable, segmentation ensembles prevail. This is because it doesn't matter how many segments you have or how the segmentation was arrived at: an individual is still scored by only one model. In consumer credit markets, for example, there will usually be different models for each product type (credit cards, unsecured loans, mortgages), whether you have a clean credit record or not (a history of paying on time or lots of missed payments), and the depth of your credit file (a lot of past credit history or very little). Likewise, in marketing different models are often constructed for different income groups, people's socio-economic grouping, new vs. repeat customers and so on.

For situations where it's all about predictive accuracy (the logic behind a dating site, or how a book retailer generates recommendations that you might like), parallel ensembles have established themselves as the method of choice. There are lots of different methods for generating parallel ensembles, but in my experience there are two that are more widely used than others.

The first method is where several different types of models are constructed using different methods. For example, you might create a decision tree using C4.5, a decision tree using CHAID, a neural network using back-propagation, a linear model constructed using logistic regression and a support vector machine – five models in total. You then create an ensemble by combining the scores together in some way. The simplest way to do this is to generate a second-stage model, where the predictor variables are the scores from the five initial models. Alternatively, you can use each model to make a separate decision about what to do with each person and take a vote. If three models predict that someone is creditworthy and two do not, you would treat the customer as creditworthy, based on the majority decision.

The second popular type of parallel ensemble is called the Random Forest.[38] This method was originally developed for use with multiple decision trees (hence the name), but can be adapted for use with other model types. The idea behind random forests is very simple. Each model is constructed using a slightly different development sample and a small set of randomly chosen predictor variables.[39] After generating many different decision trees, the results from all the individual trees are combined.

Multi-stage ensembles have received a lot of attention in academia, with methods such as boosting[40] and gradient descent.[41] However, in my experience these methods are less popular than segmented and parallel ensemble approaches in real-world environments.

There is also no reason why you can't build ensembles of ensembles if you have the time and resources available. This can improve predictive accuracy further, but results in a very complex solution.

6.9   How much benefit can I expect to get from using an ensemble?

Some people have reported the benefits of ensembles, in terms of improved accuracy, of up to 30% compared to the best single model.[42] With figures like this you might think you'd be crazy not to go down the ensemble route. Personally, I have found the typical benefits to be around 5–10% depending on the problem and type of ensemble method employed, and some observers have reported cases where the benefits of ensembles are zero or even worse than a single model.[43] The amount of improvement you can expect to see also depends on the quality of the model you are comparing the ensemble to. If you build a dodgy decision tree and compare it to a fantastic ensemble you are bound to get a huge uplift. So as with many other aspects of predictive analytics, if ensembles tickle your fancy, then you should invest some time and effort doing a little research before committing more fully.

Perhaps the most famous and informative tale of the triumph (and failure) of ensembles is the case of the Netflix prize. The movie rental firm Netflix uses predictive models to predict how its customers will rate films (i.e. how much they will like them). Netflix then targets people with offers for films it thinks they will like. In 2006 Netflix launched the Netflix prize, a three-year competition to see if anyone could develop a predictive model that was at least 10% better than their own in-house model.[44] The competition ran for three years, with a prize fund of $1m for the first person to achieve the goal (if anyone could – there was some skepticism at the time). One of the great things about this competition was that anyone could enter. You could make as many entries as you liked, and there was an online leader board to see how everyone was doing. Therefore you could try one approach, submit an entry, see how you did and then try again. The other great thing was how the competition ran right to the wire. After thousands of submissions, the 10% improvement was only achieved on the very last day of the competition, with the winning team submitting its entry just a few minutes before the competition closed in July 2009.

That's the success – a 10% improvement in the models, which has been widely reported. What gets less coverage is the fact that Netflix never implemented the winning solution. One reason was that the models were so complex that Netflix could not implement them with its existing IT, and the benefits did not justify the cost of the upgrading their infrastructure. Netflix did implement a very small part of one of the algorithms from earlier submissions that gave a 5.6% improvement in accuracy, but it was only a very small part.[45] The lesson here is that ensembles can give significant uplifts in terms of predictive accuracy, but in many real-world applications they are subject to the same operational requirements for simplicity and explicability as single model solutions, and this is their Achilles heel.

Where ensembles are used, it's most commonly segmentation ensembles that are employed. This is because the model structure remains relatively simple. There has been steady growth in the use of random forests and other types of parallel ensemble, but their use remains confined to a relatively small number of applications where model explicability is not particularly important.

6.10   The prospects for better types of predictive models in the future

New ways of building predictive models and variations on existing approaches are being developed all the time. Methods for constructing linear models and decision trees, such as linear and logistic regression, CART and so on have been around for decades, and some methods we've mentioned in this chapter originated in the 19th Century![46] Even popular ensemble methods can trace their roots back to the 1980s and 1990s. So you have to ask yourself: Are these approaches to predictive analytics a little old hat? Will they eventually be replaced by new techniques that generate significantly more accurate predictions than the models in use today?

One thing worth thinking about is the intense excitement that accompanied new developments in predictive analytics, data mining and artificial intelligence/machine learning in the past. In the mid-1980s and early 1990s methods for generating various types of neural networks were in vogue, as were other Artificial Intelligence (AI) techniques such as genetic algorithms and expert systems. In the 1990s and 2000s support vector machines and ensemble methods, such as bagging, boosting and random forests received similar attention. Recently nature inspired "swarm intelligence" approaches, such as Particle Swarm Optimization, Artificial Ant Colony Optimization and Artificial Bee Colonies have been hot topics, in particular, because these methods adopt a divide and conquer approach. They split large and complex problems into a number of more manageable chunks. Each part can be solved by a separate computer and the results then bought back together and combined. This fits very well with the growth in distributed computer systems that are increasingly being applied to Big Data.

In some circumstances a new method does provide marginal improvement over established techniques for building predictive models, or there may a very specific type of problem that a particular algorithm is well suited for. However, in practical situations the benefits are usually very small, or come with a price in terms of being very complex, or completely unintelligible to anyone without a PhD in statistics or computer science. This is not to say there aren't some problems out there that have benefited from new, cutting edge predictive analytical techniques, but in my opinion they are few and far between when it comes to predicting consumer behavior in real-world business environments.

What many people are coming to realize is that there are far better gains to be had by improving other aspects of the predictive analytics process, for example, better alignment of modeling objectives with business objectives, addressing data quality issues, obtaining new data sources, new methods of transforming data prior to model construction and improved ways for selecting the best subset of predictor variables to present to the chosen modeling algorithm. It is through improvements in these aspects of the predictive analytics process that the greatest potential benefits lie – not new types of predictive algorithm.

Please do not take this to be an entirely negative view of progress. The message I want to convey is that it is good to keep up with current developments and to look for new and better ways of doing things, but don't be taken in by the hype that always accompanies new methodologies or the pitch made by vendors with a new solution to sell. When some promising new modeling/analytical technique appears on the horizon, make sure that a proper evaluation of the method is undertaken before jumping in. Your evaluation should be undertaken independently of the supplier, i.e. *you* do the evaluation, not the people with a vested interest in seeing positive results. This should include the use of development and holdout samples, and an out-of-time sample to test the model's long-term robustness. For the test to be really sound, ideally one person will build the model using the new technique and someone else will independently carry out the evaluation using holdout sample(s) that the developer has not had access to.

Equally important is to undertake a business evaluation and a cost/benefit analysis of the process that takes into account things such as:

- Is specialist software/consultancy required to construct and implement the model?
- How much additional time, effort and cost will be required to implement the new solution within the operational environment?
- Is it important that the model is simple and/or interpretable, i.e. do you need to be able to understand how individual scores are arrived at?
- Will it be easy to monitor the performance of the model on an ongoing basis?

chapter 7

The Predictive Analytics Process

If you are going to place predictive analytics at the heart of an automated decision-making process, then you need to approach it in a controlled and systematic way. It's very tempting to just let your highly paid data scientists run wild, trusting that their analytical skills will deliver something useful. However, building a decision-making infrastructure based on predictive analytics is just like any other sort of project, whether it be building a new office building, setting up an IT data center, refitting a factory or restructuring an organization. Only a fool would hire a team of builders and let them lose on a greenfield site without an architect having drawn up the plans first, or think that all their IT problems are solved just because they have bought a truck load of cutting edge hardware.

The same goes for predictive analytics. You need to plan what you are going to do, undertake feasibility work, understand the costs and benefits and risks and issues, and have someone overseeing the whole show to ensure that it delivers what was promised on time and to budget. However, this is exactly the opposite of how a good data scientist approaches the technical part of a predictive analytics project. Exploration, experimentation, and trial and error add value. Trying different things with different bits of data, comparing alternative algorithms to see what works well and what does not, exploring a few blind alleys and so forth, actively supports the production of the model itself. It's all the other things around the model development that need careful consideration and control if predictive analytics is going to be successful.

My experience is that the data scientists themselves are not always the best people to oversee the wider task of getting models implemented and in use. That's not to say there aren't good data scientists out there who can take on board things like process reengineering, staff training and project management, but that's a bit like being both the architect and the builder.

All predictive analytics projects benefit from a degree of planning and project management. However, the approach should be proportionate to the task in hand. If you are building tactical, narrow scope, one-off type models, whose success or failure affects just a few hundred thousand dollars of revenue in a multi-billion dollar business, you probably want to adopt a light touch or Rapid Application Development (RAD) approach. As long as you have a couple of meetings with the right people and capture the decision-making process in a few e-mails then that's sufficient as far as project management is concerned. The last thing you want to do is become bogged down with a big program management philosophy and the paperwork and long delivery times that go with it.

If on the other hand, you are dealing with processes that could make or break the organization, which are responsible for millions of dollars' worth of transactions, day in, day out, or will be picked apart by industry regulators, then you many need to put n place a project management structure such as PRINCE 2;[1] i.e. something that provides a project plan, governance, accountability and a risk management framework. However, regardless of whether your project is a large or small, tactical or strategic, all predictive analytics projects should be structured in a similar way, as in illustrated in Figure 7.1

7.1 Project initiation

As shown in Figure 7.1, the very first step is to initiate the project. At the outset you need to get the right person in the organization to sponsor the project, give it their approval and get the ball rolling. For a new application of predictive analytics, or an application that's strategically important or controversial, the project sponsor should be someone senior; i.e. someone who can authorize the funds and resources required to deliver the project and who can defend the project should it be challenged by other senior managers. If the model development is more tactical or low key in nature, or you are just replacing an existing model with a more up-to-date version, then the project sponsor will probably be the head of the department deploying the model.

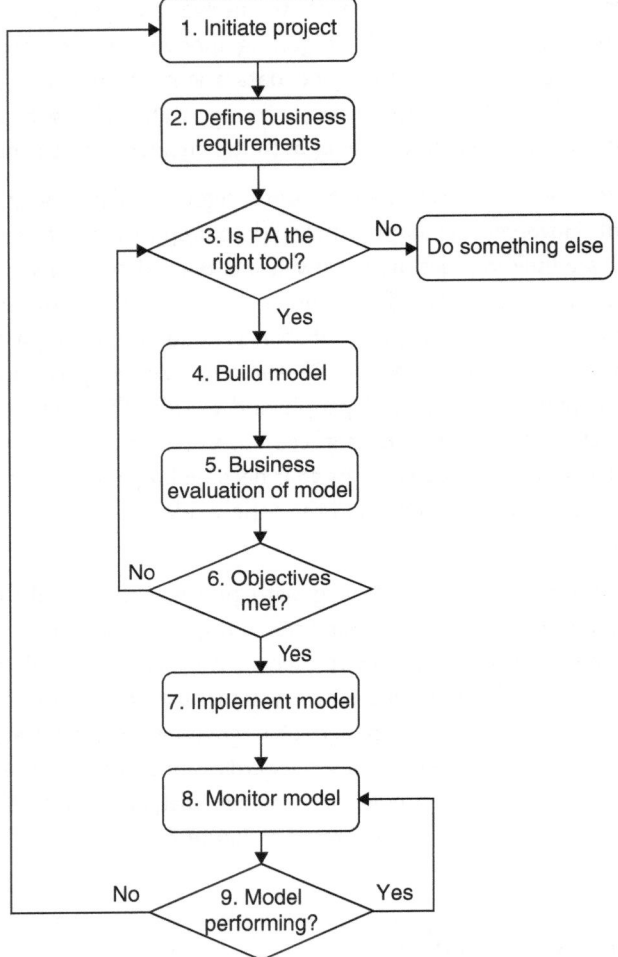

FIGURE 7.1 Process for predictive analytics

Once a project sponsor has been identified, then the rest of the project team can be assembled. This will include people with the following mix of skills and interests:

- **Project manager.** This is the central point of contact whose responsibilities include: obtaining and managing resources, ensuring company procedures are followed, facilitating communication between team members, capturing objectives, drawing up a project plan, monitoring progress against the plan and managing risks and issues throughout the project.

- **Decision makers.** These are people in authority who make the key business decisions: for example providing budget approval and signing off the model prior to implementation. The project sponsor often takes on the role of the primary decision maker, but often they will empower others to act on their behalf.
- **Data scientists.** These are the people who undertake the predictive analytics process; i.e. gathering and analyzing data and constructing the model(s). Data scientists will also be involved in defining requirements, testing, implementation and the production of documentation. The role of the data scientist is analogous to a tradesman working on a construction project or a computer programmer on an IT project.
- **Stakeholders.** Anyone affected by the model is a stakeholder. One role of the project manager is to identify stakeholders and ensure that their requirements are covered. If a new response model results in more customers phoning with questions about a product, then call center staffing levels need to be reviewed. Consequently, the call center manager is an important stakeholder. A common project management error is to neglect a business area that the project impacts upon.

This list refers to roles not people. It is not uncommon for someone to take on two (or more) roles within a project. For narrow scope tactical projects in particular, the project manager may also be the project sponsor or a senior data scientist who is given project management responsibility in addition to their other duties. A second point is that although the project manager role is central to the success of a project, project management is about the means not the ends. It is not the project manager's role to decide what the objectives are, how much the company can spend or the deadline for the project, but to deliver what has been agreed by the organization, on time and to budget. It is also a mistake to think that the project manager needs to be technically literate when it comes to predictive analytics. Yes, they do need to have an understanding of what predictive analytics is, what it is capable of and what its limitations are, but in my experience there is no correlation between a great project manager and the technical skills required to construct a predictive model.

Compared to some IT and construction projects that involve hundreds of people and cost hundreds of millions of dollars, predictive analytics projects are relatively small-scale affairs. In my experience the project team for a typical modeling project contains between 2 and 20 people, and about half of these will be stakeholders with little day-to-day involvement. Most projects involve only one or two data scientists in a full-time capacity. Even for a large program of work, involving considerable data analysis and the construction of several models, it is unusual for more than four or five data scientists to be required at any one time. If you want a consultancy to construct (but not implement or monitor) a predictive model, then expect to pay somewhere between $50,000 and $100,000 dollars per model, depending on the complexity of the data analysis involved. If you need to enhance/upgrade your IT to enable predictive models to be incorporated into your decision-making processes it will cost considerably more.

Ultimately, a project is successful because the people in authority believe that their objectives have been achieved. Typically, this is the project sponsor and/ or the decision makers. It is therefore important that everyone knows who is ultimately in charge, and everyone is attuned to what these people perceive the objectives and success criteria to be.

7.2   Project requirements

Once the project team has been established, the second step is to decide what the project is going to cover and how this is going to be done. Defining requirements for a modeling project should be approached in two stages, starting with business requirements and then feeding into technical requirements.

Business requirements describe what the business wants to do and how this will be achieved, along with the identification of any barriers that need to be overcome. Typical business requirements include:

- **Business objectives.** This is what the project is going to deliver to help the organization achieve its goals. Business objectives should be couched in terms of increased profits, improved efficiency, new capability and so on; i.e. things that are directly aligned with what the business does.
- **Scope.** What will be covered by the project and what will not.
- **Legal requirements.** In many countries there is considerable consumer legislation that organizations must comply with.
- **Implementation path.** This covers how the model will be integrated with the organization's business systems to enable it to be used on a day-to-day basis. Consideration also needs to be given to how the model will affect other areas of the business such as IT, call centers and so on.
- **Project planning.** The business will want to know how much the project is going to cost and how long it will take. Few organizations will sign up to a project without these things being agreed beforehand.

- **Risks and issues.** Risks are things that threaten the success of the project, but it is uncertain whether the risk will occur. Issues are problems that have already arisen and are obstructing progress. Action should be taken to resolve issues and mitigate risks.
- **Documentation and reporting.** This concerns how information about the project is recorded and disseminated to stakeholders. Considerable documentation may also be required to comply with internal/external audit procedures.

When discussing business objectives, it's very important to establish the baselines and success criteria against which the project will be assessed, particularly if a predictive model is being developed to replace and/or enhance an existing decision-making process. You need to agree up front how you are going to prove to people that the model does a better job than the existing process. Unless you can prove that the model is superior, the stakeholders are unlikely to give the model their approval. For example, if you build a model to improve the response from a direct marketing campaign, you need to know what the response rates/volumes are for the current direct marketing process. Otherwise, how can you say that the new model is better? Likewise, if I were building a model to replace a manual decision-making process, such as diagnosis of a particular disease, then I can only tell if the model is a success if the diagnosis rates using the model are better than the prior manual process. Just building a model that is statistically valid and predicts with a certain degree of accuracy is not sufficient in itself to prove the model's worth.

Once business requirements have been defined, they need to be translated into technical requirements that the data scientists can use to drive their approach to data analysis and model construction – and it's usually the role of the data scientist to do this. In many ways this is no different from an IT project, where a business analyst translates the organization's vision into a functional specification that the programmers can understand, or a construction project where an engineer has to translate the architect's vision into something that the builders can build.

Technical requirements predominantly focus on analytical issues. These include:

- **The modeling objective.** This is the behavior the model predicts. In statistics/data mining the modeling objective is often called the "dependent variable," the "target variable" or the "objective function."
- **The outcome period (forecast horizon).** This is the length of time in the future over which the model predicts behavior.
- **Data sources and predictor variables.** This covers where the data for the model will be sourced and which predictor variables should be considered for model construction.

- **Sample definition.** There is often lots of data that could be used for model building. Likewise, there is always data that should be excluded.[2] Sample definition is the process of deciding which sub-set of the available data should be used to construct the model.
- **Type of model and method of construction.** Sometimes, any type of model is acceptable and any algorithm can be used to derive the model. In other situations, the requirement is for a model of a particular type, constructed using a particular method.

Business requirements focus on organizational objectives, governance and control. Technical requirements cover data and analytical issues. However, it's important that business and technical requirements are aligned, and it should be possible to see how the technical requirements have been derived from the business requirements. One of the biggest risks in a predictive analytics project is not getting the integration right between the two. For example, someone may start with the technical aspects of the project and then try to retrofit some business requirements to the model they have built.

A key area in which business and technical requirements must be aligned is the definition of objectives. Business objectives need to be translated into a modeling objective before a model can be constructed. For a classification problem, the modeling objective must represent the business problem in terms of events and non-events at the level of an individual. The data scientist needs to assign a simple Yes/No flag to each case used to build the model to represent events and non-events respectively. For some problems this can seem relatively straightforward, but in reality the process contains a lot of complexity and/or ambiguity – it's not always clear what is or is not an event.

Consider a situation where we are interested in maximizing response from a direct marketing campaign for an online fashion retailer. Historical information from previous campaigns about who bought items of clothing and who did not is going to be used to construct a response model.

The obvious approach is to classify those that responded by ordering an item of clothing as a Yes (Event) and those that did not order anything as a No (Non-event). However, what about people who bought a product but subsequently returned it for a refund?[3] Do these represent a positive or negative outcome? They did respond, and if you want to take a purist view then that's clearly an event. However, what you must bear in mind is that when the business says it wants to maximize response, what it's really asking is: Did we make as much money as possible by targeting these customers? And that's what the modeling objective needs to represent.

The business will only make money if the targeted individual buys the product *and* does not return it. If you want a modeling objective that is better aligned with business objectives, then you might want to classify returns as non-responses (Non-event). There is also a third option – to treat a return as an "indeterminate event" – one that can't be classified as either a response or non-response event – it sits somewhere in the middle. These cases are then excluded from model development.[4] In this way the model is focused on predicting only clear cases over which there is no dispute. If the return cases are retained in the development sample then you might get a model that looks good when it comes to statistical measures of accuracy, but the model won't be as good at identifying the most profitable cases as a model where the return cases have been removed.

Another example where it can be difficult to define the modeling objective is consumer lending. Lenders want to lend to "Good" people who repay what they borrow and generate a profit, and decline applications from "Bad" people who default and generate a loss.[5] Let's say that a bank defines someone as in default if they are more than 90 days past due. The simplest option is to define any loan in default as an event (Bad) and everything else as a non-event (Good). Sounds straightforward? But what about cases where a borrower enters default, recovers their account and eventually pays off the debt in full? The bank will have charged the customer fees and penalty charges, and the customer will have paid extra interest. They aren't really "Bad" because the bank made a profit from them.

What about customers with a perfect repayment record, but who repaid their loans early? The bank has made a loss because the reduced interest paid on these loans didn't cover the administration costs when the loan was taken out. So these cases should be classified as bad shouldn't they? Well no, not really, because people who repay their loan early typically have a very similar profile to those that went to term and repaid their debt, i.e. "Goods." If you try to build a classification model where your population of events and non-events have very similar characteristics, then the resulting model will be a poor one. This is because the predictive analytics process relies on finding significant differences between the event and non-event populations. A better option is to remove early repayment cases all together from the predictive analytics process, to create "clear blue water" between the Goods and Bads.[6]

Another consideration when building models to predict loan repayment behavior is the forecast horizon. In theory, what banks would like to do is to take a sample of loans that were granted several years ago, classify the loans as good or bad depending on whether or not the loan was repaid at the end of the credit agreement, and then build the model using this data. However, for many types of consumer lending the repayment period is several years, and for mortgages 20 years or more is not uncommon. Obtaining full details of mortgage agreements taken out 20 years ago (credit report information and the like) is often infeasible – the data just isn't available any more. Even if data is available, things change. The factors that were predictive of mortgage default two decades ago are unlikely to be very predictive of mortgage default today.

The way lenders get around this problem is to make some assumptions and compromises when developing their models. They will take a sample of mortgages that were taken out just a couple of years ago and classify them as good or bad on the basis of their repayment behavior over the first two years. In this way the data used to build the model is far more current than if the full term of the mortgage were considered.

So again, what seemed like a simple problem (loan default) is quite nuanced when it comes to working out what the modeling objective should be, and the choice of modeling objective will have a noticeable impact on the bottom line. In consumer lending, what you tend to find is that every lender has a slightly different definition of what constitutes a "Good" or "Bad" customer and the definition also differs by product (loan, card account, auto loan, mortgage etc.). Consequently, a good data scientist will spend a considerable amount of time discussing objectives with the business to ensure that the modeling objective used to construct the model is fully aligned with the organization's business objectives. Sometimes this will involve a considerable amount of data analysis, looking at a variety of different factors, before the modeling objective is agreed. At other times a degree of subjectivity is required, with the modeling objective based on the opinions of people in the business as to what should be classified as an event or not.[7]

7.3   Is predictive analytics the right tool for the job?

Predictive analytics is just one tool that can be deployed to solve some types of business problem. It's a very good tool, but it's not always the right tool or the best tool. Once you have an idea of the requirements for the project, then that's a good time to decide whether predictive analytics is the right way to go. Reasons for deciding not to proceed with a predictive analytics project include:

- **Lack of stakeholder buy-in.** Unless the project sponsor and key stakeholders are 100% behind the use of predictive analytics, then the project is likely to fail.
- **No purpose.** There needs to be a business application for the model. There is no point building a model if no one is going to use it.
- **Lack of data.** Precictive analytics does not require "Big Data" but it does need at least a few hundred examples of the behavior you want to predict. You also need several predictor variables captured at the right point in time.
- **Deterministic outcomes.** If you can tell with certainty what someone is going to do or how they will behave, then there is no need for a predictive model.
- **Over-expectation.** The model is unlikely to deliver the promised benefits. As discussed in Chapter 1, a reasonable expectation is that predictive analytics will provide improvements in efficiency and/or bottom line measures of 20–30%. If the requirement is to generate bigger benefits than this, there is a risk that the project will under deliver.
- **Accurate, but not precise.** It's possible to build a good model, but one that is not precise enough for the job. The model may be able to discriminate between people that have a 90% chance of doing something and those that have a 10% chance. However, for the model to be useful it needs to be able to identify those with at least a 95% probability of the event.
- **No implementation route.** The model, for whatever reason, can't be implemented. Often this is due to the cost of implementation, issues of prioritization or legal barriers.

It's not unusual for some preliminary analysis to be undertaken during the business requirements phase to establish the feasibility of the project and ensure that an implementation route exists.

7.4 Model building and business evaluation

Once the technical requirements have been defined, the data scientist can begin the job of building the model (which is discussed in more detail in the next chapter). Model building encompasses a number of tasks. One is the process of extracting data from relevant data sources, loading it into the analytical environment and transforming the data into a suitable format; commonly referred to as the Extract, Load, Transform (ELT) process. They then need to apply a relevant algorithm to create the model and then carry out validation to ensure that the model works. Following validation, work is required to understand how the model impacts business processes under different usage scenarios, and measure the model's performance against the agreed performance baselines and success criteria.

Data analysis and model construction are where data scientists come into their own. Building a predictive model and evaluating it is a technical task that requires appropriate education and training. However, this does not mean that the business should not be involved. On the contrary: I would caution you to be wary of situations where a model is presented to the business as a *fait accompli*. For many applications of predictive analytics there should be several checkpoints during the model building phase, where the data scientist feeds back their interim results and findings to the business. They then refine their approach based on the feedback they have received.

The greater the potential impact of the model, and the more strategically important it is, the more important it is for the business to have clear and detailed understanding of what the data scientist has done, and for the business to ensure that the right model has been constructed to meet their requirements. For a model that is going to be used to decide how patients are treated, or will be used to make millions of dollars' worth of insurance decisions each day, I would expect there to be at least three meetings between the data scientist and stakeholders during the model-building and evaluation phases. These are:

1. **A data meeting.** This is to discuss preliminary findings about what data is predictive of behavior. At this meeting the data scientist should highlight data items that show a particularly strong, unusual or unexpected relationship with behavior.
2. **A preliminary results meeting.** At this meeting the data scientists talk about the models that they have developed. In many situations it is appropriate to discuss what predictor variables feature in the model and how much they contribute to the final score.
3. **An evaluation/usage meeting.** To discuss the cut-offs that could be employed and how this will impact on the business.

Business validation of data and the final model is important. This is because data is often inaccurate or not appropriate for use in decision making, and often this knowledge is only in the minds of business users. A piece of data may show a clear relationship with behavior, but this relationship is erroneous for one reason or another.

A classic case from financial services is the variable "Credit insurance taken out: Y/N." If you examine the relationship between this variable and the likelihood of loan default, then what you find is that on the face of it, it looks incredibly predictive. People who took out credit insurance have a much higher propensity to default on their loans than people who didn't take it out. However, what you usually find is that insurance is only sold to people after the initial decision to grant credit has been made, and it is those who only just passed the cut-off score that tend to take it up, i.e. the least creditworthy customers. Therefore it's something that cannot be used to predict default *before* the lending decision is made. If you include it in your model then the model will appear to work well when evaluated off-line, but won't perform optimally when you put it live.

Following business acceptance of the model, good practice is for the project to be written up. If the business requirements have been kept up to date, the final model documentation should be just the business requirements, plus details of the model(s) that have been constructed, how they performed and any other significant results and findings. The key test of good documentation is whether the documentation provides enough information for another data scientist, called in at a later date, to be able to replicate the model building process that has been undertaken.

7.5 Implementation

The implementation path for predictive models is something that should be considered right at the start of the project as part of the requirements stage. Many models have been delayed or never implemented due to a lack of foresight when it comes to a plan for implementing them. I remember the horror on the face of one project manager who had been expecting the data scientists to provide a mechanism that would deliver scores into the business, when all he got was an Excel file detailing the structure of the model. Another time I recall a group of data scientists ranting at a software vendor because the model implementation software that they had purchased only implemented linear models and decision trees. They had spent months developing a complex neural network solution and had no way of implementing it. This was not the vendor's fault: the model developers should have asked the right questions at

the requirements stage. This is a great example of data scientists running free, without considering the constraints that they had to work within.

If a model is for a one-shot or *ad hoc* purpose, implementation may mean an additional ELT process to extract details of the people you need scores for, and these scores are generated using the same analytical server and software used to construct the model. The scored file is then handed over to whoever in the business needs it. If we are talking about a response model that is going to be used to select cases for a mail shot (similar to the Booles case study in Chapter 2), those with high response scores are selected and their details loaded into the organization's customer contact system. The marketing department then provides the content and formatting of the mail shot, and the contact system prints it and dispatches the mail shot to those that have been selected.

This type of approach is fine when one is using models infrequently, in an *ad hoc* type of way. If, on the other hand, the model is going to be used to score people as part of an operational system, then the model needs to be integrated into that system. This is to allow the scores to be calculated as and when they are needed, without any requirement for people to be involved in the scoring process.

In the early days of predictive analytics, the implementation of a model within an operational system such as an airline booking system or a loan application system would have been undertaken as an IT project. A programmer would code up the logic for calculating the model scores as a sub-routine, using popular programming languages of the time such as Cobol or Fortran. Whenever a score was required the sub-routine would be called by the system that required it. In theory, there is nothing wrong with this approach and many organizations still implement their models in this way. However, what one tends to find is that this approach is slow and error-prone. The project will need to be put into the IT work queue and it may be months before the work gets done. It is also very common for the programmer to write logically correct code that works, but which contains errors in terms of the assigned scores. For example, coding 10 as 100 or +27 instead of –27.

Another problem is when it comes to upgrading the model or changing the cut-off rules and other decision logic around the use of the model. If the marketing department wanted to change the cut-off of a response model to increase the number of people being targeted, or the credit department wanted to increase the credit line for some people, then it might take weeks or even months to get the necessary IT resource to change the relevant code

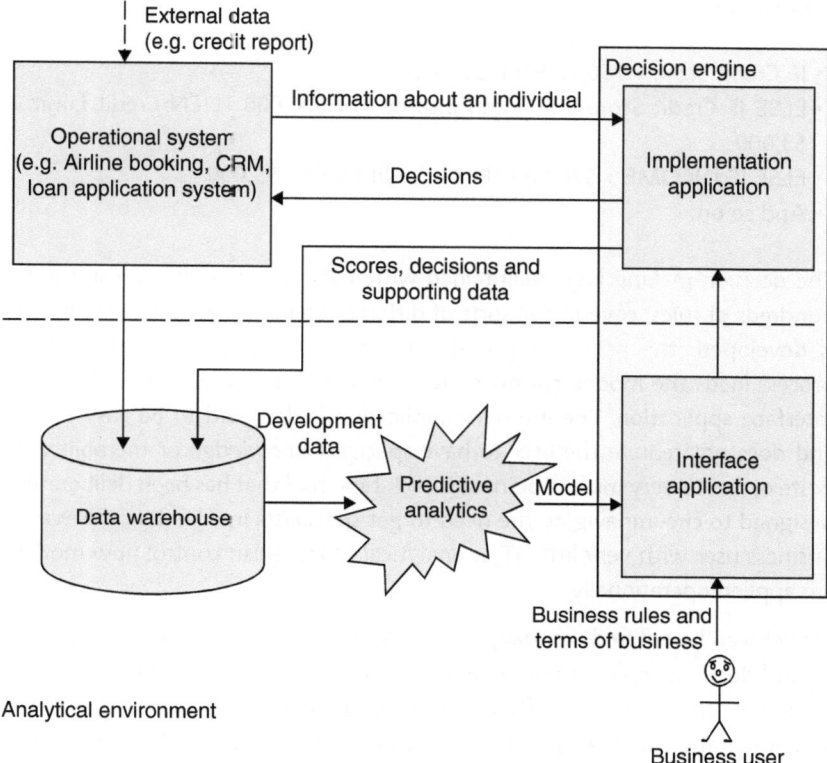

Operational environment

External data
(e.g. credit report)

Decision engine

Operational system
(e.g. Airline booking, CRM,
loan application system)

Information about an individual

Implementation
application

Decisions

Scores, decisions and
supporting data

Development
data

Data warehouse

Predictive
analytics

Model

Interface
application

Business rules and
terms of business

Analytical environment

Business user

FIGURE 7.2 / **Decision engine**

module. To overcome these problems many organizations now implement
their models using decision engine technology, as shown in Figure 7.2.

Some operational systems have decision engine capability built in. In other
situations the decision engine is purchased as a standalone piece of software.
A decision engine usually contains two parts. The first part, the interface
application, sits on a PC/server which is not directly connected to the
operational environment. The interface application is where the model, cut-
off rules, override rules and terms of business are maintained. It is important
to appreciate that the interface application captures much more than just the
logic for calculating scores. It will contain the entire decision-making logic
around the use of the model. A decision engine that is used to assess credit
card applications, for example, will allow the user to specify all the terms of
business for the product, such as the credit line (credit limit), the interest

rate, when the card expires and so on. This is done using IF THEN type logic. For example:

- IF Credit Score < 750 THEN DECLINE
- ELSE IF Credit Score < 800 and Income < = $70,000 THEN Credit Limit = $3,000
- ELSE IF INCOME > $70,000 THEN Credit Limit = $5,000
- And so on.

The decision-making logic maintained within the decision engine can run to hundreds of rules, covering all sorts of different scenarios. When a new model is developed, the person responsible for maintaining the decision-making process loads the model, cut-off scores and associated decision rules into the interface application. The interface application is designed to be easy to use and does not require the user to have specialist knowledge or the ability to write code. It's very much a point and click type tool that has been deliberately designed to circumnavigate the need to get specialists involved. In this way a business user with very little IT or analytical training can control how models are applied operationally.

Models can be entered manually into a decision engine, but a key strength of a good decision engine is that it allows models to be automatically transferred to it from the analytical software used to build the model. This speeds up implementation and reduces (although not entirely removes) the potential for human error.[8]

Predictive Model Markup Language (PMML) is a set of protocols that standardize the representation of many common model types. If the analytical software used to construct the model and the decision engine used to implement the model both support PMML, then this allows models to be automatically transferred between the analytical environment and the operational environment, even if one vendor's software is used to construct the model and a different vendor's software is used to implement it.

Once all the models, cut-off rules, override rules and terms of business have been set up in the interface application, testing functionality allows the user to pass samples of data through the decision engine on their PC to check that scores are being calculated correctly, to see how many people hit various override rules, and the expected impact of the chosen cut-off strategies. Many decision engines also include test and learn capabilities. This allows the user to explore the business impact of different models, cut-offs and decision rules in order to optimize the full decision making process. Once the user is happy with the decision-making logic it is uploaded to the implementation module, where final testing occurs prior to go live.

Whenever a decision is required, the operational system calls the implementation module. The implementation module calculates scores, applies cut-offs and business rules, and returns decision codes that specify what action should be taken. So for our credit card example, the decision engine not only supplies a score, but also a decision code (accept or decline the application) plus the credit limit, APR and any other information required to open the account for those that have been accepted. These decisions are then used by the operational system to drive the actions that are taken in relation to the application. For those that have been declined, a communication will be dispatched explaining the decision. For those that have had their application for a credit card accepted, the terms of business are used to generate a credit agreement that is then sent to the customer to be signed, and so on.

7.6 Monitoring and redevelopment

The performance of predictive models decays over time. There comes a point when a model needs to be replaced with a new, more up-to-date version. Some models last for years before a new model is required. Others decay very quickly and need to be redeveloped on a much more frequent basis. If a predictive model is being used for a tactical one-off purpose, then model decay is probably not that much of an issue, but you should still evaluate how well the model performed and incorporate what you learn into future projects.

If you are going to be using the model every day to drive a business-critical decision-making processes, then model decay is a very important issue indeed. Model decay is often a neglected issue, but is one of the most important aspects of model implementation. "A model is for life, not just Christmas" is a mantra I repeat to my clients, who are often more focused on getting predictive models in place, rather than developing long-term plans to deal with what comes after implementation.

With regard to model decay, the questions I get asked time and time again are:

- "How long will our model(s) last?"
- "With all this data changing so rapidly, surely my model will decay very quickly?"

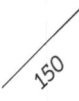

- "As the speed of change increases won't I need to redevelop more often?"
- "How will I know when it's the right time to redevelop?"

My first response to these types of question is: "Don't get confused between data change and relationship change." It's only when the relationships between the predictor variables and behavior change significantly that a model loses its ability to predict. Just because data has high velocity and/or volatility does not in itself mean that a new model is required more frequently than in a stable, slow-moving environment where data changes less often.

To give an example, if one day all cases scored by your response model are young people, and then next day it's all middle-aged people, this doesn't necessarily mean that the relationship between age and response has changed – only that the numbers of people with different ages have altered. Maybe yesterday's marketing activity was targeted at university students (aged 18–22) while today the focus is on parents with teenage children (aged 35–55). If the two age groups were both considered when the model was originally constructed then that should be sufficient to enable the model to treat both age ranges appropriately. A new model is not required.

Let's look at another example. Consider a model that predicts the risk of heart attack in the next six months. The model is based on patients' medical records and uses data such as blood pressure, age, gender, alcohol consumption, cholesterol levels and so on to make its predictions. When the patient visits their doctor, their information is updated and a revised risk prediction is generated for the next six months. A model that predicts heart attack should take a very long time to decay, and still be working well many years after it was developed. Why? Because human physiology doesn't change much. The factors that predicted heart attacks ten years ago are probably the same today. So as long as those factors were included in the model when it was originally built, the model will remain optimal.

Now imagine that instead of visiting their doctor every six months, all the patients have a biometric sensor implanted in their arm that measures the same things that the doctor did, but dynamically refreshes the data every few seconds. Suddenly there is orders of magnitude more data available than before. Having the data available more frequently does not in any way invalidate the existing model. It will continue to work as well as it ever did. However, the extra data provided by the sensor creates some exciting new possibilities. One is that the model can be used much more dynamically. Model predictions can be updated in real time, and if the risk predicted by the model rises very dramatically the doctor can contact the patient, rather than waiting for them to come in for their next appointment. You could even implement the model as a phone app, linked to the sensor, which alerts the patient of a potential problem if their risk of a heart attack exceeds a certain threshold. If you take this idea to its logical conclusion then you could program the phone to automatically call an ambulance in expectation of the heart attack that is about to happen.

As a rule, a predictive model only needs to be redeveloped in one of four circumstances:

1. **New/better data becomes available.** The additional data allows more predictive models to be constructed.
2. **Relationship change.** There is a paradigm shift in how individuals behave or the environment they inhabit that results in a significant change in the relationships between the predictor data and the outcome being predicted.
3. **Data scarcity.** When data that the model needs to calculate scores becomes unavailable.
4. **When the organization changes its objectives.** The model no longer predicts what the organization wants.

With regard to new data, the more timely data provided by the sensor enables new predictor variables to be considered. You might find that large jumps/falls in blood pressure in a very short time, or the difference between daytime and nighttime readings, are important predictors, i.e. things that the doctor could never have measured without the sensor. The existing model still works as well as it ever did, but it may be possible to build a much better model using the new predictor variables that are now available.

As to the second point, I've argued that the heart attack model should last for a long time, but that's not to say the model is immortal. Suppose that all adults in the population are now given statins to reduce their cholesterol levels – a proposal that has received serious consideration in several countries.[9] Consequently, almost no one in the population has high cholesterol any more, and therefore cholesterol level is something that no longer needs to feature in the model.

The heart attack scenario demonstrates a relatively stable and slow-moving environment. In other situations things are much more dynamic. This is particularly true when there is interaction between fast moving-data streams. One example is the take-up of financial services products from online comparison sites. What's important is the relative differences between the offers available, which change all the time throughout the day. A model that was very good at predicting product take-up first thing in the morning may be out of date by the afternoon.

Another fast-moving area is fraud. Fraud is somewhat unique in the predictive analytics world because it is the one behavior that people deliberately seek to mask in order to avoid detection. You don't see much evidence of this in other areas, like healthcare or marketing. I've never heard of anyone changing their day-to-day activities in an attempt to outwit direct marketing campaigns! Fraudsters will use a given behavioral pattern for as long as it works, and once detected will change tack. In the "good old days," before the Internet, when a fraudster found a new way to deceive it would take a while to filter out into the wider criminal community and what they found often didn't go far, staying a small-scale local affair. These days, new scams can be worldwide in minutes.

One common type of fraud is to seek funds from a bank using a stolen identity, knowing that the bank only carries out detailed identify checks when the funds requested are more than, say, $2,500. The fraudster will have tried different amounts, had some requests refused, some accepted, swapped information with other fraudsters and come to the conclusion that the bank is good for a hit of between $2,400 and $2,500. Once the bank has had a few hits like this it will react to the problem. Maybe it responds by setting a lower value threshold for performing detailed identity checks, or by changing its verification procedures for low-value withdrawals. In response, the fraudsters will change their behavior, adapting to the new checks and procedures that the bank has put in place – it's a continual game of cat and mouse. What this means is that fraud models need to be redeveloped more frequently than many other types of predictive model used in financial services.

Another reason for redeveloping a predictive model is when data used by the model is no longer available. Sometimes an operational process changes, and as a consequence that data is no longer captured within the organization's systems. The marketing team may decide that asking insurance applicants ten questions about their lifestyle is too many, and this must be reduced to eight, so that there is less incentive for them to abandon their purchase part way through. If the discarded information features in the organization's predictive models, then those models will no longer be able to calculate accurate predictions and will require redevelopment. Similarly, sometimes changes in the law make it illegal to use certain types of data, or require changes to the way data is gathered and used. This is what happened in the insurance industry in the EU, when in 2011 it became illegal to differentiate insurance premiums on the basis of gender, which is one of the most predictive items of data for many types of insurance. Consequently, most insurers had to rebuild their models with gender excluded.

The final reason for redeveloping a predictive model is when the organization is no longer interested in predicting that behavior. If we continue with the heart attacks scenario, maybe the doctor decides that, given that the patients all have a sensor that is taking biometric readings on a real-time basis, what is really needed is a model that predicts if someone is going to have a heart attack in the next 24 hours. This will allow them to be admitted straight to hospital, before the heart attack occurs, based on the model's predictions. The six-month forecast is therefore no longer required.[10]

How can you tell when it's time to redevelop a model? The usual way is to instigate a monitoring regime that determines what, if any, decay in model performance has occurred since the model was originally developed. At regular intervals the performance of the model is assessed and compared with the model's performance when it was originally constructed, and also with previous monitoring results to allow trends in model performance to be observed.

For traditional applications of predictive analytics, such as credit scoring models and actuarial models used to predict life expectancy, monitoring often occurs on a monthly or quarterly cycle. Each monitoring period a data scientist spends a few hours poring over various reports to see where and how the performance of the model has changed, and it's often just as important to understand why a model is no longer predictive as it is to know that it isn't working in the first place. In more dynamic environments monitoring needs to occur more frequently, and will often be driven by automated alert systems. A data scientist only gets involved when the model's performance moves outside of agreed tolerances.

A current trend is to remove the data scientist all together once an initial model has been constructed, and automate the monitoring and redevelopment process. If model performance falls outside of agreed tolerances then the parameters of the model are automatically adjusted to bring it back into line.

This approach has some merits, but is only really suitable for a small minority of environments where data quality is guaranteed, the data feeding the model is relatively stable, and there is no requirement to evolve the model beyond its original design parameters. In the vast majority of cases, the risks and limitations of an automated update process for predictive models does not make it a viable option – any short-term efficiency savings are outweighed by longer term risks and lost opportunity costs. However, having the monitoring system automatically generate models that the data scientist can examine and compare to the live model is something that does have its uses and is a practice that I recommend.

7.7   How long should a predictive analytics project take?

How long is a piece of string? For a new application of predictive analytics, the time and effort required to plan, build and implement the model will depend on the scope and context of what you are trying to do and the industry one is working in, along with the cultural approach to project management, risk and regulatory oversight. Once you have predictive analytics in place, then redeveloping the model for the *n*th time is a much less time-consuming process than doing it for the first time. All the infrastructure should be in place and all you may be doing is replacing one set of parameters with another,[11] which can be done very quickly.

To try to put this into perspective let's consider some different scenarios. Hopefully, this will be of some practical use when it comes to planning your own projects. We'll start by thinking about something close to the ideal. Imagine you have a good data scientist working for you who is fully conversant with the data that is going to be used to build the model. The data is near perfect and only a few hundred megabytes in size. Maybe you already have a great data set that has been used to build a response model, a revenue model and so on, which are about to be used to target people as part of a marketing campaign. Just before the campaign goes out, somebody says,

> *Hey, any chance you can just knock me up another model that predicts the expected response time – i.e. how long after we target someone can we expect a response? We can then align staffing levels at our call center with the expected response pattern to maximize call center efficiency.*

The response time model does not need to be implemented as such. All the data scientist needs to do once the model is finished is to score up their model development sample, and use it to produce a table or graph showing the response profile over time. This is passed to the call center manager to aid with planning their staff roster. So something like:

- On day 1 we expect to receive 20 calls
- On day 2 we expect to receive 140 calls

• On day 3 we expect to receive 490 calls
• On day 22 we expect to receive 7 calls
• On day 23 we expect call volumes to drop to zero.

That's it. The model has served its purpose and can be discarded. Given this "ideal" scenario, a good data scientist should be able to produce a workable model in a few hours, including a relevant (1–2 page) summary report that show the expected response profile over time with confidence limits and other relevant metrics.

In a similar vein, a developing trend is towards the use of in-database analytics. Analytical tools are incorporated into the fabric of the organization's data warehouse, making it unnecessary to extract data and load it onto a separate analytical server. This functionality was originally conceived as a mechanism for speeding up the production of standard business reporting, but vendors are increasingly putting forward their in-database analytics systems as a way to speed up the production of predictive models. If all you are doing is refreshing the parameters of an existing model that isn't particularly business critical, then again we could be talking about hours, rather than days or weeks, to redevelop the model.

That's the best case. Now let's consider something at the other end of the spectrum. Imagine that you are now working in a large bureaucratic government department that is developing predictive models for the first time. They are going to use predictive analytics to identify benefits claimants that are most likely to make an incorrect claim. These people can then be targeted with tailored help/support before they make a claim, which will reduce the incidence of error further down the line and generate a huge cost saving.[12] Being government there will probably be a big project management culture and lots of hoops to jump through. It may take anything up to six months before everyone is on board, the stakeholders have been identified and the business requirements produced. Data is spread all over the place and there is little in the way of explanation as to what the data means, or of its validity or stability. Lots of people understand their own little part of the organization's data, but no one understands it all. In this type of scenario, expect to spend 3–4 months gathering, extracting and preparing the data and another 2–3 months to analyze it, understand it and build some models. There is then another 3–6 month period to load and test the models within the benefits system before the models are finally put live. So there is a total end-to-end time of between 14 and 18 months, with 5–7 months to build the model once the business requirements have been agreed.

Take another example. In retail banking, the vast majority of consumer lending decisions are made using predictive models. A bank will have rebuilt its credit scoring models many times, and the types of data that feature in its models will be fairly consistent from one generation of the model to the next. However, to redevelop a credit scoring model you are probably looking at anything from 3–12 months from the day the project kicks off to when the model finally goes live. A huge amount of this time will be spent on documentation and validation to satisfy auditors and regulators that the models are stable and robust, and that you understand the effect that the model will have on the bank's capital position. Only a few days or weeks will actually be spent on data analysis and model construction. Move to the bank's marketing department and many of the audit/regulatory requirements don't apply. Consequently, models tend to be built and implemented in days or weeks rather than months.

A common mistake when discussing timescales for predictive analytics projects is to confuse:

1. The time required to develop a model once you have the data in a nice, clean, well prepared format.
2. The time required to gather and prepare data and then develop a model.
3. The time required for the full end-to-end project lifecycle, from project initiation through to model implementation.

If all your data is in perfect condition, nicely formatted and so forth (Scenario 1), then in theory you can build a model in just a few minutes – it's just a case of hitting the right button in your software and then casting your eye over the model diagnostics. However, if you have to go through a large data-gathering exercise, where you need to clean and prepare the data (Scenario 2) then that could easily take 2–3 months. A full end-to-end project (Scenario 3) can often take six months or more.

It is very common to see these differences when a business is talking to an external developer. A standard quote from a solutions supplier is that it will take 2–3 months to deliver a model.[13] That's true, but they are usually only talking about Scenario 2, which covers the data analysis and model building phase after you have initiated the project, gathered data and supplied it to them. Once they have built the model you then have to implement it. Therefore be careful when comparing the timescales quoted by internal and external developers – make sure that both quotes cover the same range of deliverables.

chapter 8

How to Build a Predictive Model

Once you've decided what your objectives and timescales are, who needs to be involved, and how you are going to implement and use the model, the data scientist can get on with the task of building the model. Figure 8.1 (which expands upon Step 4 in Figure 7.1) shows the steps required.

Model building is an iterative process. A lot of the work involved in building a predictive model is about getting the best out of the data and then squaring this with business requirements and any constraints that exist (such as limitations on what can be implemented operationally and legally). A data scientist may build dozens of interim models during this process before a final model is decided upon.

In describing the model-building process I'm not going to provide any code or formulas. This is partly because it's not that sort of book, but also because there is loads of good specialist training out there if you need it, lots of Internet resources and many great books if you want to get into the nitty-gritty (see Appendix B for some references). Instead, I'm going to take you through the *principles* that data scientists should follow when constructing a model to ensure that it predicts well and satisfies any business constraints.

You don't need to be a data scientist to get involved in the model-building process. Likewise, if a data scientist has any wits about them, then they will always make sure that they identify key business people and draw upon their knowledge of data and systems to enhance what they do.

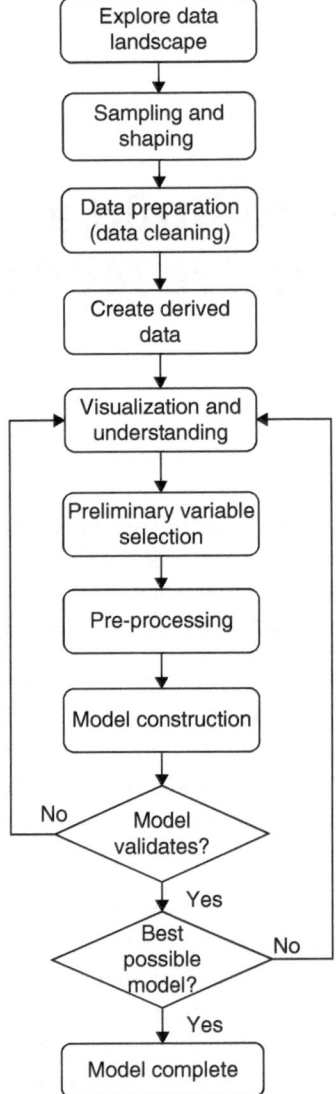

FIGURE 8.1 Model construction process

8.1 Exploring the data landscape

One of the first tasks that a data scientist needs to undertake is understand what data is "out there" and what data can be considered for use in the predictive analytics process. In an ideal world an organization will have detailed documentation (metadata) describing all of the data it holds, what it means, how it's formatted and where it comes from. In practice, metadata is patchy. Some data is well documented, but a lot of information is only held in the minds of people that use it on a day-to-day basis. It's also unfortunate that many organizations hold data that they no longer understand, but they are afraid to get rid of it just in case it's something important that a business-critical process relies upon. There is also the issue of new data sources, both internal and external, that could be considered in a predictive model, but are not well understood because the organization has never looked at them before. What this means is that a good deal of effort is required to explore the data landscape from which data will be gathered. Some avenues to explore include:

* **Existing metadata.** One should seek out any documentation that exists about the data that an organization holds.
* **Current decision-making systems.** If data is already used for decision making or features in existing predictive models or other types of prediction system, then there is a good chance that it will also prove useful for your problem.
* **People working in operational areas.** These people know what types of data the organization uses when dealing with people on a day-to-day basis, and will have a view about what data is important.
* **External data sources.** For example: credit reference data, web pages and so on.
* **Internet search/peer discussion.** Try to find out what people have done for similar problems in the past, and then try to match that back to what is available to you.

If time/resources are short and if there is a lot of data to consider, think about things in terms of which data sources are most likely to provide information about primary, secondary and tertiary behaviors and so on (as discussed in Chapter 4) and prioritize accordingly.

8.2 Sampling and shaping the development sample

It's very rare for a predictive model to be built using information about every individual that you know about. For many problems, only a relatively small sub-set (a sample) of available cases are suitable. Consequently, one undertakes sampling to create the development sample that is used for model construction. There are three reasons why sampling occurs:

1. **To create a representative development sample.** The data used to create a predictive model should be representative of the operational environment. Otherwise the model won't work very well. If historical data exists for people of all ages, but the model is only going to be applied to people aged over 30, the model development sample should only contain people aged over 30.
2. **To create a recent and timely development sample.** Relationships between data and behavior change over time. The more recent the data used to build a model, the more likely it is that the relationships captured by the model still hold true when the model is deployed. Therefore only recent cases are included in the model development sample. Older cases are excluded.
3. **To create a manageable development sample.** Historically, computers did not have the storage and processing capabilities to analyze populations containing millions of cases. Consequently, common practice was to take a sample: say 1 in 10 or 1 in 100 of the available cases. The capabilities of modern computers mean that much larger samples can now be taken, but there are still times when sampling is advisable for this reason.

Take the case of a car manufacturer wanting to develop a classification model to predict which potential customers are likely to make a claim against the car's warranty within 24 months of buying the vehicle. They want to do this because they can then offer bigger discounts to customers who are unlikely to claim (who get a low score).

The ideal situation would be for the manufacturer to gather predictor data about people who bought cars exactly 24 months ago on one particular day, and see which of them claimed under warranty in the 24 months following their purchase. In this way they would have the most up-to-date sample possible when building their model.

One problem with this sampling strategy is that even popular makes of car don't sell many units in a single day. In the UK, sales of the bestselling model average only about 300 a day.[1] Of these, only a minority of buyers, perhaps 1 in 4 (25%), claim under warranty in the first two years. As discussed in Chapter 4, 75 warranty claims is far too few to construct a predictive model. You need at least several hundred, and ideally several thousand, cases of each behavior (claim and no claim).

To overcome the numbers issue, the manufacturer doesn't just take cases from one day but looks at sales over a longer sample period (sample window). What an established manufacturer, such as Ford or BMW, could do in theory is look at warranty claims going back decades. However, taking a sample period that is too long also has its problems. This is because cars (and driving habits) have evolved considerably over the years. The reliability of new cars today is much better than it was in the 1980s or 1990s. In addition, some of the things that can go wrong in a car today just didn't exist back then. Air-conditioning, central locking, start-stop engine systems, dynamic suspension, electronic engine management systems and so on didn't exist or were confined to a few high-end brands. Likewise, people in the UK had different attitudes to things like speeding and drink driving. If data about all these old warranties was used to build the model, then the model would be a poor one because it wouldn't be representative of cars and their drivers today.

Based on 300 sales a day and a warranty claim rate of 25%, the manufacturer would probably have enough cases to build a model using one month's data. However, car sales are seasonal. There are peaks at certain times of the year, reflecting different buying patterns and behavior. Therefore taking car sales data from just one month isn't ideal either. What would make more sense is to gather data over a period of a year or so to cover the seasonality issue, i.e. the model development sample contains examples of sales made between two and three years ago. This is a bit of a compromise in terms of the recency of the data, but it provides a large stable sample that should allow some reasonable models to be constructed. These sorts of trade-offs, between recency, sample size and the representativeness of the sample, are something that need due consideration by the data scientist.

Having decided upon the sample window, further shaping of the sample is required. In particular, cases in the development sample that won't be representative of the operational environment need to be excluded. Typical model exclusion reasons include:

- **Out of context cases.** The population contains cases that aren't relevant to the problem at hand. Perhaps the manufacturer's discount strategy is only going to be applied to personal buyers. However, the manufacturer sells a considerable number of fleet cars to companies, which involves a completely different pricing model. Therefore all company car sales are excluded from the model development sample.
- **Change of customer populations.** Certain types of people that the organization dealt with in the past won't be assessed by the predictive

model going forward. Perhaps the manufacturer used to target first-time drivers, but no longer offers finance to this group. Therefore there is no point including first-time drivers in the sample population.

- **One-off events.** If some behaviors can be attributed to specific one-off events that won't occur again, then it is prudent to exclude these cases. If one batch of cars all had a specific fault due to an error in the manufacturing process then it would be prudent for the manufacturer to exclude these cases from their model development sample.

8.3 Data preparation (data cleaning)

Data is dirty, filthy, messy stuff. Often it's incorrect, missing or badly formatted, particularly where humans have been involved in creating and/or collecting it. Sometimes numeric data is held as text, or text data is forced into fixed-length fields resulting in some data being truncated, and so on. Consequently, a lot of the time and effort involved in a predictive analytics project can be spent "cleaning" the data before it's ready to be used. The main data cleaning tasks include:

- **Coding missing values**. Data is often missing for some cases. Maybe family details are collected by the online version of a form, but the questions are not asked on the paper version. The fact that data is missing and the reasons why it is missing can be useful predictors. Good practice is to assign default values to missing data, and different default values if the data is missing for different reasons. A ratio variable, for example, may be "missing" because the denominator is missing, the numerator is missing, both the numerator and denominator is missing or the denominator is zero. Each type of missing tells you something different.
- **Consolidation**. Maybe a field holds different values to mean the same thing. For some people, marital status is described by "M," "D" and "S" for married, divorced and single respectively, but for others these attributes are coded "1," "2," "3." Work is required to convert all the cases of "1" to "M," every "2" to "D" and so on.[2] Similarly, data may be spread across different records which need to be merged together. If a credit card is lost or stolen, standard practice is to freeze the account and create a new one. To get a complete view of the customer you need to merge information from the old and new accounts.
- **Standardization**. Good practice is to have all data of a similar type in a similar format. If time living at an address is held in months and time in

current employment held in years, then consider changing the format of one of the variables.

The above list applies mainly to structured data such as customer account records, but similar data cleaning issues are also relevant for unstructured data such as text and images. Consequently, a whole range of procedures have been developed to clean up this type of data. For example, some data cleaning tasks that one would apply to natural language (letters, web pages, phone transcripts and so on) include: correcting spellings, removing punctuation and stop words (e.g. "an," "it," "is," "the"), and "stemming" to standardize word endings (e.g. replacing "started," "starting," "starts" with "start").

8.4 Creating derived data

Whatever data you have, wherever it has come from, whether it's structured or unstructured, data in its raw form is not often very useful for prediction. Usually, there is far greater merit in deriving new types of data from it rather than using it as it is. There are several types of derived data that one should consider when constructing predictive models:

- **Ratio variables.** Height and weight are useful for predicting certain medical conditions, but Body Mass Index (BMI)[3] is more predictive than either weight or height in most cases. The balance and spending limit on a credit card are interesting data items, but when it comes to predicting future spend, the ratio of card balance to credit limit is always a better predictor than card balance and card limit on their own.
- **Accumulated variables.** This is where you begin with data items of a similar type, which are combined together in some way, for example by taking the sum, average, minimum or maximum. A good example is the items on a shopping list. You can sum up the prices to calculate total spend, calculate the average spend per item, the cheapest item, the most expensive, the ratio of cheapest to most expensive item and so on.
- **Time-based variables.** It is common to have data about the timing of specific events. Some events are a one-off, such as date of birth, which is used to derive age. Other events occur more than once. If you know when someone shopped in your store, then you can calculate measures such as time since last shop, average time between shops, number of times they shopped in the last three months and so on.
- **Extracted variables.** Language, text, pictures and video often contain useful information, but cannot be used directly to build a predictive model.

First, it needs to be transformed into a numeric or categorical format. If you are trying to predict financial crime, then you may want to create counters for the number of times key words and phrases are used in e-mails, such as: "fraud," "off-shore" and "off the books."

For many projects there are potentially millions of derived variables that could be considered – far more than any analytical system can deal with. Therefore an important part of the model-building phase is using business knowledge to come up with ideas as to what types of ratio, accumulated, time-based and extracted variables one should consider.

8.5   Understanding the data

A golden rule of predictive analytics is that you shouldn't build models using data that you don't understand or about which there are doubts concerning its source, stability or validity. This is particularly relevant for models that are going to be used for many months or years as part of an operational decision-making system. From a user perspective, if a data scientist presents you with a model and can't explain what the data is or where it comes from, then that's a cause for concern. It's therefore important for a data scientist to examine:

- **The distribution of data**. Does the spread of data conform to common sense and business expectation? If "Number of children" is a predictor variable that takes values in the range 100–1,000, with an average value of 252, then something is clearly amiss, either in the data itself or the metadata that describes it.
- **How well the data is populated.** If you only know the residential status for 1% of people, and its missing for the other 99% – then what does this tell you about the relationship between residential status and behavior, if anything?[4]
- **The relationship between the predictor variable and the modeling objective.** If I'm building a model to predict heart attacks, and my data shows a negative correlation between blood pressure and heart attack (high blood pressure = low chance of heart attack) then that's a relationship that goes against all other medical evidence and therefore must be treated with caution.

In all these cases, understanding the nature of the data is a key task for the project team. Sometimes it will be easy to explain away something unusual. If I have a variable called Monthly Household Income that has typical values such as 456556, 176434 and 1000045 then that's suspect because the values are far too high to be true. However, this may be because the data is stored without a decimal place – the real values are $4565.56, $1764.34, $10000.45, and maybe someone who works with this data on a regular basis can confirm if my assumption is true, in which case that's fine. In other cases the reasoning might be more subtle and a data scientist may need to do extensive "forensic" work to determine exactly why the data has the values it does.

What can cause real problems are those rare times when the data throws up something genuine that is really strange, surprising or new. The initial reaction will be that the data is suspect and should be excluded from further analysis, but this is the wrong action to take. A great example of this type of issue was a credit scoring project that I worked on with a retailer that provided credit to people with very poor credit histories, who couldn't get credit elsewhere. In every credit scoring project I or other members of the project team had ever worked on, the variable "Number of County Court Judgments"[5] was massively predictive. If you had been taken to court for an unpaid debt in the past, then this was hugely predictive of future repayment default and the more court cases against you and the more recent they were, the worse the credit risk the individual represented.

However, this wasn't the case for this project. Having a County Court Judgment indicated good repayment behavior. This seemed crazy. However, when we explored the problem we realized that the relationship was a genuine feature of the customer population. The retailer was operating in the most sub-prime of sub-prime credit markets, where almost everyone had a history of bad credit, and default. One reason that a court judgment was a positive indicator was that it acted as a proof of identity – the person was who they said they were. In this type of sub-prime market there are a lot of opportunist fraudsters who submit fake applications in the hope of making a quick buck at the lender's expense. A judgment also demonstrated at least an ability to get credit (even if they didn't repay it!). Those without a Country Court Judgment were so uncreditworthy that they had no recent credit history at all.

8.6   Preliminary variable selection (data reduction)

If you only have a few hundred potential predictor variables, then it's probably worthwhile examining the relationship between each one and behavior in turn, using appropriate visualization tools (graphs and/or tables) and discussing what you find with the project stakeholders. However, as the richness of datasets increases, with many thousands of potential predictor variables becoming the norm, it's just not feasible to explore everything manually.

Many popular methods of model construction also struggle to create models in realistic time if presented with more than a few hundred variables, even if the number of cases in the sample is relatively small (a few tens of thousands). Therefore, once you have an analytical dataset containing all of the potential predictor variables, it's usual to do some preliminary work to reduce this to a manageable sub-set before further analysis and model construction are undertaken.

There are two types of tool that are used to perform variable selection:

- **Filters.** These examine each predictor variable in turn. A numerical measure is calculated, representing the strength of the correlation between the predictor variable and the modeling objective. Only predictor variables where the correlation measure exceeds a given threshold are retained.[6]
- **Wrappers.** A wrapper takes a group of predictor variables and considers the "value add" of each variable compared to other variables in the group. If two variables tell you more or less the same thing (e.g. age and date of birth) then one will be discarded because it adds no value. Stepwise linear regression and principal component analysis are two popular wrapper methods.

Filters are the easiest, quickest and most commonly applied approach to variable selection, but they do have some limitations. In particular, they don't consider correlation between predictor variables. As discussed previously, variables such as Age and Date of Birth are perfectly correlated, and therefore you don't need both in a model. However, a filter will consider each on its own. If Age is ranked highly by the filter, Date of Birth will also rank highly.

Wrapper approaches are often superior to filters, but take considerably more time and effort to apply. One way to apply a wrapper is to do it as a two-stage process. First, you take small sub-sets of predictor variables (maybe 50–100 at a time) and examine each sub-set in turn. All of the variables that are retained from each sub-set are then examined together by the second stage wrapper.

8.7   Pre-processing (data transformation)

By this stage in the model-building process there should a nice clean, well-formatted development sample which contains:

- **All potential predictor variables.** These are the data items that are going to be considered for inclusion in the predictive model.

• **The modeling objective.** This is a clear representation of the behavior that we want to predict. For a classification problem this will be a binary indicator (0/1 or Y/N). For regression problems this will be the quantity that you want to predict.

If unstructured data such as text or images have been considered for inclusion, then suitable text/image analytics will have been applied to extract the useful bits, so as to create a suitable structured representation of that data.

If we are talking about text, then this might be predictor variables which represent the number of times certain words and phrases appear in documents and/or flags indicating if the text captures positive or negative opinions (sentiments). If we are talking about images, then predictor variables will exist to represent key features that have been identified from that image. For example, if we are talking about medical imaging (scans and X-rays), then there will be variables to indicate unusual features, such as scarring, odd lumps or textual differences in tissue. If we are talking about images from social network sites, then you might have indicators for things like gender, estimated age, eye color and so on that have been generated from face analysis/recognition software.

By this stage all of this data will be in the form of a single table (a dataset), similar to that shown in Table 8.1.

Table 8.1 relates to the warranty claim model. As well as standard data items to allow the data to be matched to other sources (the two leftmost columns), the development sample also contains a whole host of predictor variables that will be considered for inclusion in the model. The two rightmost columns contain behavioral (outcome) data. Given that this is a classification model, Time to Claim has been used to derive a 0/1 indicator variable that will be used as the modeling objective.

The final step before pressing the button on the modeling software is pre-processing (transformation) of the predictor variables. Different model construction techniques display varying degrees of sensitivity to the way data is presented to them and many work better if pre-processing is undertaken to achieve the following:

• **Linearization.** Transformations are applied so that the relationships between the predictor variables and behavior are (approximately) linear. Having linear relationships is important for methods such as linear regression and logistic regression. If the relationships in the data are highly non-linear

Table 8.1 The final data set

Identity data		Predictor data						Behavior (outcome) data	
Ref.	Purchase ID	Driver age (years)	Income ($000)	Residential. status*	Number of speeding convictions	…	Annual mileage last year	Time to claim (months)**	Modeling objective***
1	079345	42	39	T	0	…	5,250	–1	0
2	079995	62	119	O	0	…	11,190	–1	0
3	108765	29	82	O	0	…	27,000	17	1
4	109045	55	164	O	0	…	16,050	9	1
5	109755	49	44	T	3	…	2,290	–1	0
…	…	…	…	…	…	…	…	…	…
9,998	196607	19	33	X	0	…	16,940	23	1
9,999	196712	49	63	L	1	…	5,220	–1	0
10,000	108766	31	58	O	0	…	8,270	11	1

* O = Owner, T = Tenant, L = Living with parent, X = Unknown/other.

** –1 indicates that no warranty claim was made within 24 months of purchase.

*** 1 = Warranty claim within 24 months, 0 = no warranty claim within 24 months.

then poor models will result when using these methods. Linearization is less important for techniques such as CART and neural networks, but linearizing the data often improves the performance of these methods also.

- **Standardization.** If one predictor variable takes values in the range 100 to 1,000,000 and another takes values in the range 0.001 to 1, then the model weights will be very different, even if the two variables contribute equally to the model. It is good practice to transform numeric variables (such as income and age) so that they all take values that lie on the same scale.

The easiest way to pre-process data, which linearizes and standardizes the data, is binning and the use of indicator (dummy) variables to represent each bin – as discussed in Chapter 6.

Two common questions that arise during the binning process are:

1. **How many bins should there be for each variable?** Too many and the model will not be robust and over-fitting is likely. Too few and the resulting loss in the granularity of the data will reduce the predictive performance of the model.
2. **Where should the boundaries between bins lie?** For example, is it better to create an indicator for ages 18–24, 18–25 or 18–26?

With regard to the first question, one rule of thumb is that each interval should contain at least 100 cases in total (and a minimum of 30 of each class for classification problems), but ideally each bin should contain several hundred cases to minimize over-fitting. Another observation is that in many cases just a few bins usually works very well – sometimes just two or three is sufficient. Even with very large data sets, containing millions of cases, there is often little benefit to having more than 20–30 bins to represent numeric variables such as age or income.

With regard to the second question, the easiest approach is to create equally sized bins, each containing the same number of observations, and this will give good levels of performance. However, there are a number of algorithms that can be applied to optimize bin definitions in line with your modeling objectives.

For many problems in predictive analytics it also makes sense to define bins to align with common sense and business knowledge. So you might run an automated binning algorithm to start with, and then manually tweak the intervals that are created. If you are looking at the variable "age," then 65 is a common retirement age and retirement is a life event that affects many behaviors. Therefore, it makes sense to ensure that 65 is an interval boundary.

Similarly, for a credit card, negative and zero month-end balances often indicate very different card usage behavior than a positive balance. Therefore whatever the automated binning procedures tell you, it makes sense to consider splitting out negative and zero balances.

Binning and the use of indicator variables is the most popular data pre-processing approach, but there are others.[7] One is to use the bins to define the weights of evidence variables.[8] Other pre-processing methods involve functional transforms (log and power functions) and the use of curve fitting (splines). However, in general these methods don't significantly outperform indicator based models.

8.8   Model construction (modeling)

Model construction is where the magic occurs. A data scientist will set up their specialist analytics software, point it at the data and let it crunch the numbers. These days most analytical software has many automated features and there is no need for users to do any calculation themselves. There are even cloud-based tools, such as Google Predictive API, that will build you a model for free, as long as you have some data to upload to it.[9]

Many packages also provide facilities that simplify the cleaning and preparation of the data, and have the functionality to create derived variables and pre-process the data to create indicator variables. So in some senses you don't need a degree in math or statistics, or be a whizz programmer, to build a predictive model. However, the software needs to be controlled and the outputs interpreted in the right way, and this is where the knowledge and experience of a good data scientist is vital if you want an optimal model, rather than a "will do" model.

An understanding of the mathematics underpinning the various algorithms used to create predictive models isn't essential, but having this knowledge gives a data scientist important insights into why the software has created the model that it has and allows them to fine tune the model to improve its performance. In particular, if the model isn't right in some way, then the data scientist will know which parameters of the software to tinker with that will hopefully correct the problem and deliver a better quality model.

There are all sorts of things that the data scientist will do with the analytical software, but the main tasks that are required to set up the software include:

• Specifying which variable captures the behavior to be predicted (represents the modeling objective).

- Specifying which predictor variables to consider in the model and which to exclude.
- Deciding which type of predictive model to construct and the method that should be used to generate it. For example: construct a linear model using minimized least squares, or a decision tree using CHAID.
- The algorithm's parameters. Most algorithms have several parameters that control how they generate a model. It takes experience and a lot of trial and error to find the best parameters for any given problem.

As part of the model construction process, the software will generate all sorts of outputs describing the evolution of the model, together with various statistics and standard measures of performance. The data scientists will then evaluate the outputs. If they think that the model is a good one, they will move on to the validation stage. If not, they will tweak the parameters of the software, maybe try a different type of model or algorithm, and try again.

8.9 Validation

A major problem when building predictive models is that it's quite easy to find relationships that are the result of random patterns in the development sample, but which don't exist in the wider population. The result is that if you measure the performance of the model using the development sample the results will be over-optimistic. They will not be a true representation of how the model performs when presented with new cases when the model is deployed. This problem is called over-fitting.

To assess the true performance of a predictive model and to determine if over-fitting has occurred, the model needs to be tested on an independent sample of data that was not used to construct the model. Good practice is for a data scientist to have two samples when building a model. The development sample is used to construct the model and the validation sample is used to test it. By comparing the model's performance on the development and validation samples it is possible to determine if over-fitting has occurred and by how much. A small amount of over-fitting is quite common (i.e. slightly worse performance is observed on the validation sample), but this is not necessarily a problem if the model still performs well. However, if the degree of over-fitting is large and the model's performance is poor, the model developer will need to go back to the drawing board and build a new model using a different set of parameters.

Model building is an exploratory process. Lots of different models get built and tested using the same validation sample. As a result, after many cycles of model building and testing, a degree of over-fitting to the validation sample is likely to occur. Therefore best practice is to maintain a third sample – a holdout sample – that is used to perform the final analysis of model performance. Any improvement figures, results, estimated benefits and so on, that are reported back to business stakeholders should always be based on the holdout sample – otherwise one runs the risk of over-stating the benefits case for the model.

For an operational model with a long forecast horizon (such as credit scoring models), it's also good practice to take additional holdout samples from different time points. These "out of time" samples are used to test the long-term stability of the model's performance. This is important because the performance of most predictive models decays over time. It is quite probable that a predictive model developed on a sample of data that is several months or years old won't perform quite as well as initially indicated from validation and holdout samples taken from the same time period as the development sample. If some deterioration in performance is observed this does not necessarily mean that the model is useless, but the benefits case needs to be revised downwards to reflect any deterioration that is observed.

8.10   Selling models into the business

After a model has been built and validated, that's not the end of the story. Often the stakeholders need to be convinced that the model is fit for purpose before they'll authorize its use. How much do non-technical "business people" need to know about the predictive models that have been developed for them? If all we are talking about is a regular update to refine an existing model then maybe very little, but somebody somewhere will want to know what impact the new model will have on the bottom line before it goes live. Similarly, not many people may need to know about a marketing model used to target people with an offer, but the marketing director will still want to want to know how many extra sales will result from using the new model.

At the other end of the spectrum, if the model is the first of its kind to be developed for your organization, and you are trying to convince a group of skeptics that the model is sensible and makes predictions in a logical way, then you probably want to know quite a lot about the model in order to be able to win them over. Likewise, if you are staking your professional reputation on implementing predictive analytics then you'd be crazy not to demand to know something about the model structure, how it performs, what type of individuals tend to get high scores and which get low scores. The last thing you want to do is implement what seems to be a great model only to wake up one morning and find some press story about how your organization mainly targets women or never deals with people from certain ethnic groups. This is not because the model deliberately allocates certain people high or low scores, but because of some indirect effect that creates a statistically valid, but socially unacceptable, bias in the model.

Just like human experts, all models have strengths and weaknesses and it makes sense to know what these are. Otherwise, you put yourself and your organization at risk. Consider a model that gives higher scores to people with higher incomes. This will mean that on average women will score lower than men even if gender is not in the model.[10] Is that a problem? If it's a model used to set insurance premiums in EU countries then you would be acting illegally if you used this model without addressing the gender issue. On the other hand, if the model predicts response to direct marketing activity for a particular make of car, then that's probably not an issue at all.

For a large-scale modeling project, where the model is going to be implemented within a strategically important operational decision-making system, then at the very least the stakeholders need to be given some sort of information pack and/or presentation about the model and how it performs. The data scientist(s) may have done all the hard work of developing the model, but it's up to you (the business) to decide how to use it, i.e. what type of cut-off strategies should be employed to meet your business objectives, what level of risk you are willing to accept and so on.

Often there won't be one single best cut-off. Instead, there are a range of options allowing trade-offs between different costs and benefits. For an insurance claim model, one cut-off may result in the same number of new customers, but a 15% reduction in claims. Another cut-off may mean that you can increase the number of profitable customers by 10%, but there will be a corresponding 10% rise in the value of claims, and so on. A data scientist should not aim to provide you with every detail of the project in minute detail, but they should be able to demonstrate to you that the project has delivered what was intended, and present the options for implementation in a way that is easy for you to understand, and hence make an informed decision about how the model should be deployed.

Standard practice for communicating model results to non-technical business stakeholders is for a meeting/presentation to be held once the model has been

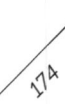

developed, but prior to model implementation. At this meeting, the data scientist needs to explain to relevant stakeholders what they have done and provide assurance as to the quality of the deliverables. In particular, they need to demonstrate that the objectives and success criteria specified at the start of the project have been achieved (or if not, why not). This is to enable the stakeholder group to decide whether or not the finished model can be signed off and put to use.

There is much advice that can be given about presentation style and format, but most things that lead to a good presentation are based on common sense principles. Perhaps the most important thing to bear in mind is: what do the people in the audience want to hear? The data scientist should put themselves in the audience's shoes. They should focus on the key points that they need to get across and avoid digressing into minor technical details or irrelevant subject matter that result in a loss of audience attention. Following a KISS (Keep It Simple Stupid) philosophy isn't a bad idea.

If a data scientist is dealing with people with different backgrounds and experience, then they should identify who is important and tailor their material to those people's objectives and level of understanding. If they don't know someone's background, then it is generally a good thing to err on the side of caution and pitch to the intelligent (but naïve) layperson. The data scientist should not assume any technical knowledge, but should not talk down to the audience either. Unless dealing with an exclusively technical audience, a good data scientist will leave <u>all</u> formulas, equations and acronyms out of the discussion. However, they should have at their disposal more detailed technical material that they can call upon if necessary. A key thing that a data scientist should bear in mind is that someone is unlikely to judge something a success if they don't understand what's been achieved or whether or not their objectives have been met.

I remember (with some discomfort) sitting with a group of senior managers and listening to a presentation from a data scientist who was very bright but had little experience of talking about predictive models outside of their peer group. Consequently they were very naïve when it came to presenting models to a non-technical audience. The data scientist opened their presentation with the immortal lines of: *"Following the application of a filter approach to variable reduction, we generated tree ensembles using our chosen variable subset and a maximized entropy measure ..."* to an audience with no background in model development. The poor analyst was almost laughed out of the room before their more senior colleagues managed to pull things back from the brink.[11]

What a good data scientist remembers when presenting models is that the goal isn't to show off their technical expertise or to bamboozle the audience with buzzwords and clever terminology, but to convey an understanding of what has been achieved in terms that the audience understands. If, as more often than not, the primary goal of the project is to reduce costs, increase profits and/or increase the size of the customer base, then this is what the primary focus should be.

My own preferred method of proceeding in such situations is to have three levels to the material I present. To begin, I present a short (two to three minute) summary, using a single slide/sheet of paper. This contains only the most important information about the project, what has been achieved and how this meets the business objectives and success criteria specified in the project requirements. For a credit scoring project to develop a model to assess the creditworthiness of new loan applications, the summary may be presented as a few bullet points stating the benefits that implementation of the model will deliver. For example:

Implementation of the new model will deliver the following benefits:

- An annual increase in annual net profit of between $5m and $6m.
- An annual reduction in bad debt write-off of between $0.9m and $1.1m per annum.
- No significant change to the volume of new loans granted.
- A 15–20% reduction in underwriter referrals.

After the summary I cover other things that are important from a business perspective: for example what the model looks like if it is a linear model or a decision tree, and when the new model can be expected to be implemented. Risks and issues around implementation and the schedule for monitoring the model once it goes live are also covered if necessary, as are impacts to the operational areas of the business. Finally, I'll have a technical appendix included with the slide deck or as an attached document covering some of the more specialist aspects of the project in detail. Usually this won't be presented in full, but will be used to address any technical questions about the project that are raised.

8.11   The rise of the regulator

In the good old "Wild West" days of predictive analytics only those who had some direct interest in the model (i.e. the project stakeholders) concerned themselves with the quality of the model and the impact it would have on their part of the organization. This was great from the data scientists' perspective because it meant that they could spend their time on all the interesting analytical stuff and not spend too much time worrying about boring things like documentation and audit trails. However, as predictive analytics becomes increasingly important to what organizations do and has ever more influence on peoples' daily lives, so governments, industry regulators and auditors are increasingly turning their eyes towards the predictive models that organizations employ.

One reason for the increasing interest in predictive models is the risks to an organization should they get their models wrong. Imagine if an insurer accidently inverted all the points in a claim propensity model. As a consequence, very low insurance premiums would be offered to very high risk people and vice versa. The impact would be disastrous for the insurer, given that once a policy had been taken out it can't be rescinded. Another concern is the impact of model usage on private individuals, in particular, the risk that an automated decision-making process creates some sort of bias, resulting in unfair treatment for some people. How do I know that organizations aren't using illicit things like gender, race, religion and so on in their decision-making processes?

Government interest in the dangers of predictive analytics dates back to the 1970s, when the USA enacted the Equal Credit Opportunity Act 1974, forbidding the use of certain types of data in credit scoring models. The Act also gave individuals the right to demand to know the reasons why their credit applications were declined, i.e. the factors in the credit scoring models that led to them getting a low credit score.

In the 2000s, international banking regulators encouraged financial services organizations to use their credit scoring models as key inputs to their capital calculations, as required by the international BASEL II/III capital accords.[12] As a consequence, banking regulators expect credit scoring models to be built to a high standard and to be rigorously validated and tested before being put to use. I know of several UK banks that have teams of more than 20 predictive analytics experts whose sole role is model validation. These people don't do any data analysis or model building themselves – all they do is question, challenge and evaluate models built by other people in the organization.

This requirement for validation of predictive models by independent third parties is not confined to banking. The International Solvency II insurance regulations are designed to bring stability to the insurance industry. As a result, many types of insurance models are subject to similar levels of scrutiny as those in banking. Government audit functions are increasingly interested in the workings of predictive models that are used to protect billions of dollars' worth of tax revenues. Likewise, any health authority that uses predictive analytics to make decisions about patient care, or to prioritize waiting lists, needs to be absolutely sure that those models are working as intended and do not display bias.

As predictive analytics becomes ever more pervasive, then one can expect greater scrutiny to be applied to predictive models used in all sorts of applications, and particularly those areas, such as healthcare, criminal investigation and national security, where the consequences of getting a model wrong could be very serious indeed.

This is not to say that all predictive models in all industry sectors will end up being vetted to the same standards or subject to the same level of scrutiny. A model to target customers for a particular brand of car will never be subject to the same audit requirements as capital models employed by the major banks. However, internal/external audit requirements are something that should always be discussed as part of the business requirements process, lest you risk being tripped up later down the line when your models are found wanting.

Unlike business stakeholders, when discussing models with regulators and auditors the expectation should be that the audience is technically expert and will want to understand in detail the technical aspects of model construction. A regulator who is doing their job will challenge many of the decisions made about the way the project was undertaken and want to understand the decision-making process that led to the models being developed in the way that they were. One therefore needs to be well prepared prior to such meetings. In financial services, many organizations spend days, if not weeks, preparing for a regulatory review meeting that might last just an hour or two.

Some typical pitfalls that an organization can fall into when presenting their models to regulators include:

- **Over emphasis on financial benefits.** A regulator has no interest in how much the model adds to your bottom line. What they want to explore is the risks the model exposes the organization to and/or the impact on the wider public.

- **Production of separate documentation just for the auditor/ regulator.** Auditors want to see an organization's internal documentation that they use themselves because it gives them additional insight into that organization's approach to predictive analytics.
- **Poor understanding of data.** Building models without understanding the nature of the data that supports the model is a potential risk. You should be able to explain where the data has come from, what it means and what (if any) issues there are with it.
- **An incomplete strategy for implementation and monitoring.** Building a model is one thing, using it is another. An organization needs to assure regulators that rigorous testing of the model implementation has occurred, that scores are correctly calculated, and that post-implementation a monitoring regime will be in place to monitor the performance of the model on an ongoing basis.
- **Insufficient senior management understanding.** Regulators in some industries expect senior managers to have an understanding of the models employed by their organizations. This doesn't mean managers need a degree in statistics, but they should know why a model was constructed, how it is used and any risk or issues associated with it. Regulators take a dim view of senior managers who shirk their responsibilities by delegating regulatory responsibilities to junior members of staff.

No model is perfect. Given more time, more money, better data and so on, a superior model can always be built. Regulators know this and do not expect every model to be perfect. However, what they do want to see is evidence that a logical and well thought-out approach to the predictive analytics process has been adopted. This includes consideration of the risks that any weaknesses in the model creates, the impact of those risks if they are realized, and the organization's plans to mitigate those risks should they occur. Good practice is to include a section on the model's strengths and weaknesses within project documentation to show that you have thought about them.

9

Text Mining and Social Network Analysis

Text mining and social network analysis have both come to prominence in conjunction with increasing interest in Big Data. Both deal in large quantities of data, much of it unstructured, and a lot of the potential added value of Big Data comes from applying these two data analysis methods. We shall begin by discussing some of the ways in which text mining can be applied to predictive analytics, before moving on to discuss social network analysis.

9.1 Text mining

All of the methods used to create predictive models require data to be well structured, and the data must be categorical (e.g. occupation, marital status and gender) or numeric (e.g. age, income and time at address). A predictive model can't be built if the data is not in one of these two formats. At first sight this seems to rule out the use of written correspondence, the spoken word and so on, which is highly unstructured and does not appear to be numeric or categorical in nature.

Until the early 2000s this wasn't really a problem. Organizations didn't have the facilities to store large amounts of natural language data (speech, letters, magazine articles, scientific papers, web pages and other types of verbal/written information) in electronic format. Even the most advanced organizations were still focusing their efforts on getting more value out of well-formatted numerical/categorical data held in their account management and customer relationship management (CRM) systems. The idea of applying analytical techniques to this type of unstructured data was very much the last thing on people's minds.

These days, things are very different. Organizations are awash with textual and other types of unstructured data. All sorts of customer correspondence, which until the late 1990s would have been held in racks of filing cabinets, is now stored in electronic format; and then there is all that Internet data, such as tweets, blogs and web pages, to think about. So it would be a real shame if we couldn't do something with all that data to infer things about people and improve the quality of the predictive models that are available.

Text mining (text analytics) is the process of extracting information from natural language and distilling it into a structured format that can then be used for analytical purposes. A lot of text mining activity is backward looking. It's about understanding what has just happened so that a response can be formulated to it, or to improve an organization's response the next time something similar occurs. Governments, for example, want to monitor Twitter and other popular media following a policy announcement to gauge public opinion. If the policy is proving very popular then they will seek to maximize the media attention that it gets over the next few hours and days, but if the public doesn't like it then they will initiate damage limitation, and/or try to focus the public on something else.

In the past, a government would have had to wait until the evening news reports/morning papers to obtain an initial opinion of how the policy was being received, and the only practical way to get a representative view of voter opinion would have been to carry out a survey, asking a relatively small sample of people (a few hundred) their views on the policy via face-to-face or phone-based interviews. It may well have taken several days to carry out the survey, and then several more days to analyze and interpret the results. These days, reactions to political announcements can be monitored in real time as people post their views online, meaning that the politicians can respond to trending public sentiments in minutes rather than hours or days.

Marketing departments are also interested in text analytics to gauge customer opinions about their products and services, reactions to new product launches, the impact of price changes and so forth. Likewise, security services and law enforcement agencies have huge lists of suspects that they want to monitor, but they just don't have the time or resources to carry out detailed surveillance on everyone. Text analytics is one tool they use (amongst others), to scan the e-mails, texts, phone calls and websites of those on their suspect lists, looking for indicators of criminal/terrorist behavior. This allows them to narrow their focus to a few specific individuals. In a similar manner, internal fraud teams apply automated text analytics to internal e-mails to identify potentially fraudulent behavior by employees. On a slightly more upbeat note, the same techniques can be used to flag up topics/issues that are causing employees problems in their work, which can then be addressed. If several people in different departments independently comment on the "slow server today" then the IT department can be notified that there might be a problem before it is formally reported to the IT help desk.

9.2   Using text analytics to create predictor variables

When it comes to predictive analytics, the same text mining tools that are used to identify views or sentiments expressed in documents can also be used to extract relevant predictor data from those documents and represent it in a well-structured format (going forward we will use "document" as a general catch-all term to mean any type of natural language, whatever its nature or source). The procedures that create predictive models can then use the structured data that has been extracted. Typically, the predictor variables will be of one of two types:

- **Within document predictors.** These indicate that a document contains certain features – for example, yes/no flags to indicate if the text expresses positive or negative sentiments about the subject under discussion. Similarly, flags are created to indicate whether certain words or phrases are explicitly mentioned, or to identify what the topic of discussion is within the document – for example, to identify which of an organization's products a document refers to.
- **Across document predictors.** These provide measures of correlation/similarity with other documents. If an e-mail contains very similar text to other e-mails that are known to have come from fraudsters, then this is probably a good indication that the e-mail is also fraudulent. Likewise, documents can be grouped based on the topics to which they refer.

9.3   Within document predictors

One way to create predictor variables from a set of documents is to use indicators to represent how often individual words or phrases appear within each document. Let's imagine that I am an industry regulator looking to identify individuals who may be carrying out illicit financial transactions within

their organizations. In addition to all the standard stuff about employees, such as their pay grade, how long they have been working for the company, what department they work in, any previous misdemeanors and so forth, I also have access to all e-mails sent to the employees of the company. Imagine that one of the e-mails contains the following text:

> After transferring the assets to the Bahamas, I waited until the audit had been completed before moving them back to the UK. Later that month I shipped the assets back to the Bahamas again. I think I managed to move them without any issues.

So what I could then do is to create predictor variables for each word as follows:

Word	Variable Name	Variable value (Word count)
After	Text1	1
transferring	Text2	1
the	Text3	5
Assets	Text4	2
To	Text5	3
Etc.		

One thing to remember is that we are dealing with large numbers of people, words and e-mails. If your model development sample contains details of 10,000 employees, and on average there are 1,000 emails from each employee, then that's ten million documents. All those documents will contain tens of thousands of different words. If you adopt the approach of creating a predictor variable for each unique word, then what you find is that you need thousands of predictor variables: one to represent each word that appears in any of the documents. You will also find that many of the words appear in only a few documents and therefore the predictor variables that represent those words are set to zero most of the time. This is not a very efficient approach to text analytics, and life would be a lot easier if the problem could be simplified in some way. Thankfully, in most cases you don't actually need to consider every single word. Many words add very little information value to a document and therefore there is no point considering them. Likewise, similar words are used to mean the same thing. To help with this process, a standard pre-processing step applied during text analysis is "normalization." Typical normalization involves:

- **Removal of stop words.** A stop word is text that appears very frequently in prose, but adds very little to the overall meaning. There is no definitive list of stop words in any given language, but common examples of English stop words include: "The," "a," "is," "to," "of," "as," "and," "which," "on," and "that." A typical stop word list contains around 1–200 words.[1]

• **Stemming.** This is the process of removing the inflection from words, replacing them with their root. For example, "complete," "completed," "completes" and "completing" have "complet" as their common root. Some stemming procedures use simple rules to identify the root, such as removing "ed," "ing," "s" and "es" from all words in the document.[2] Other approaches use look-up tables or probabilistic algorithms to infer what the probable root is.

• **Standardization**. In many languages there are several words that all mean the same thing. For example: "I transferred the assets," "I moved the assets" and "I conveyed the assets" are all different ways of saying the same thing. Sometimes it also makes sense to replace pronouns such as "it" and "them" with the object to which they refer. "Tell Simon if you see him again" becomes "Tell Simon if you see Simon again."

So when you put these things together, the previous sentence is normalized to something like:

> *Transf assets Bahamas wait audit complet transf assets UK later month transf assets Bahamas think manag transf assets without issues.*

Normalization has reduced the number of different words from 30 to 13. The result is not easy reading, but you can just about infer the gist of it from the normalized form. Some information has been lost, but not that much, and the data is much more amenable to common text analytics procedures in this normalized form.

Following normalization, further words can be discarded if they appear very infrequently. If you cast your mind back to Chapter 4, I stated that one typically needs at least 30–50 cases to be able to say with confidence if a predictor variable is predictive of behavior. So what this means is that if a word appears only a few times in your development sample, there is no point creating an indicator variable for it. By removing these rarely used words (say where they feature in fewer than 30 documents) the number of predictor variables can be substantially reduced.

An alternative approach is to start with a list of target words/phrases and see which if any of them feature in the documents. If I was interested in identifying tax evasion, I would create a predictor variable to indicate if any well-known tax havens, such as "The Bahamas," "The Cayman Islands" and so on were mentioned. This approach works well if you are dealing with a well-defined domain and you know what words are likely to be of interest. Often there will be only a few dozen key words or phrases that you need to scan documents for. Therefore you only need that number of predictor variables – which makes the problem much more manageable.

9.4   Sentiment analysis

A development on basic text analysis is Sentiment Analysis. Sentiments are people's thoughts, opinions and feelings about things. Admiration, approval, affinity, pity, hatred, disdain and so on are all sentiments. In an ideal world one would seek to identify precisely which sentiments were being expressed in a given document, and these could then be used to generate predictor variables. However, most forms of sentiment analysis are relatively crude in this respect. They simply attempt to classify people's feelings in terms of whether they represent a "positive" or "negative" view of the subject about which the sentiment is being expressed. Consequently, most approaches to sentiment analysis would not make any distinction between admiration, approval and affinity (positive sentiments) or hate, disdain and pity (negative sentiments).

A common approach to sentiment analysis is to combine word counts with a lexicon of words/phrases associated with positive and negative sentiments. Within the lexicon, each word is assigned a positive or negative rating. So words/phrases such as "Nice," "Loved it," "Perfect," "Fantastic" and "Best" will have a positive rating, whereas words such as "Despise," "Hate," "Terrible," "Bad" and "Nasty" have a negative rating. In a typical system, words might get a rating from +10 to indicate a very strong positive association, to –10 for a very strong negative association. Words such as "Adore" and "Despise" would get very high positive and negative rating respectively, whereas less emotive words such as "Nice" and "Inadequate" would get lower ratings.

To create an overall sentiment rating for a document, the individual ratings for all the words from the lexicon that appear in the document are added together. The overall sentiment rating is then used as a predictor variable within the predictive analytics process.

Sentiments don't exist in isolation. A person expresses a sentiment about something. This can be yourself, a product, another person, the government, a press article or a whole host of other things. Often, as part of sentiment analysis, one needs to identify both the sentiment and the thing to which the sentiment refers. Sometimes this is obvious. For example, if a product website accepts customer feedback, then it is reasonable to assume that if someone provides a comment then it's going to be about that particular product. However, if one is analyzing more general text, work needs to be done to determine what the subject of the document is.

Named entity extraction is a methodology used to determine who, what or where is being discussed within a document, i.e. the subject(s) under discussion. It is possible to determine with a high degree of certainty which words in a document describe subjects, and which do not. Some approaches use probabilistic methods, based on a large database of similar documents. Others consider the structure of sentences and paragraphs in order to pick out the relevant subject(s). This leads to additional predictor variables to indicate whether a document refers to a particular subject or not, together with the sentiment that is being expressed about that subject. So in a complex document there may be several subjects and several sentiments.

One application of named entity extraction is in the security services. They will scan e-mails, texts, blogs and so on, seeking documents about certain subjects – such as terrorist events or other matters of national security. Once these have been identified, sentiment analysis is applied to try to determine whether those topics are being discussed in a positive or negative manner.

9.5 Across document predictors

As well as looking for features within documents, such as the frequency with which words appear, the subject matter or the sentiments being expressed, a very popular and successful text analytics approach is correlation analysis. The starting point for correlation analysis is to have a set of benchmark documents that are already labeled to indicate whether they do or don't relate to the behavior you are interested in predicting. For instance, with the fraud example, the benchmark set would be created using e-mails that are known to have been associated with previous fraud cases involving asset transfers, together with examples that are known to be completely innocent. An example of what this might look like what is shown in Table 9.1.

Even if the Fraud/Not fraud labels in Table 9.1 had been kept hidden, you would probably have been able to have a good guess at which ones referred to fraudulent transactions and which did not. In doing so you would have picked out phrases that seem similar to the e-mail in question. For example, a common theme in three of the fraudulent examples is the movement of assets. The other common theme that applies to the fraudulent cases is secrecy

Table 9.1   Class labels for documents

Benchmark text	Behavior
I have successfully moved the assets abroad without anyone knowing.	Fraud
Whatever you do, don't say anything to the finance department, if they get even a whiff of what's going on they will be all over us like a rash.	Fraud
I can think of three possible locations to move them to. The Bahamas, the Cayman Islands and Cyprus.	Fraud
I have deleted all correspondence and no one should ever be able to trace the asset transfer back to us.	Fraud
As long as we keep things quiet we should be able to make a lot of money from this deal.	Fraud
I'll pop in and see you tomorrow. Are you free around 10?	Not fraud
Have you seen Bob's new haircut – it's like something from the 1970s!	Not fraud
Can I remind everyone that we operate a clear desk policy.	Not fraud
Sorry, but I can't make the meeting because I'll be on holiday.	Not fraud
I think that we need to resolve the problem by the end of day.	Not fraud

and the withholding of information. Algorithms that perform correlation analysis look at how similar each document in the development sample are to those in the benchmark set in terms of word usage and/or sentence structure.[3] The procedure allocates each document in the model development sample a correlation measure ranging from zero to one (low to high correlation). The correlation measure is then used as a predictor variable.

Another approach is clustering: grouping together documents that have similar features in terms of their structure, the words they contain and the subjects to which they refer. This is very similar, in theory, to the clustering approaches discussed in Chapter 6. An attractive feature of document clustering is that it can be done without reference to the behavior that one is trying to predict. Documents are assigned to clusters purely on the basis of their similarity to other documents within those clusters. After assigning documents to clusters, the cluster number is then used as a predictor variable in the predictive analytics process.

9.6   Social network analysis

Network analysis is the study of how groups of things are connected and the interactions that occur between those things. The road network that links cities together, electricity grids that connect generators and consumers, and the wiring between the transistors on a microprocessor are all networks. Network analysis is important, because by understanding the nature of a network you can anticipate the impact of changes to that network, such as what happens to traffic flow if a new connection is made between two cities (a new road is built between them), the impact on consumers if a power line goes down (a connection is broken), or the impact on performance of alternative wiring schemas between the transistors of a microprocessor.

Social network analysis relates specifically to entities and the relationships (connections) between them. It can be applied to all sorts of entities, including people, companies, countries, academic institutions, charities and so on, but for our purposes we shall limit the discussion to the relationships between individuals. It's also worth pointing out that although people often talk about social networks and Internet services such as Facebook in the same breath, social network analysis is about our relationships in general – it's not confined to social network sites or online interactions.

My family is one network I'm part of. I also have a network of peers via LinkedIn, fellow employees at work and my friends that I socialize with. There are also criminal networks, support networks, customer and supplier networks and so on. These networks are also interconnected with each other. I can be a relation, friend, peer, criminal and so on all at the same time, with some connections appearing in several networks, while others are unique to just one.

So how would someone go about using information about my family, friend or peer group network? One application is political campaigning. Politicians have only limited resources at their disposal, and want to concentrate their campaign efforts where they are likely to have greatest impact. They don't want to waste their time on people who are solid Republicans or staunch Democrats and who will always vote for their favored party – they want to identify the floating (swing) voters who are open to persuasion.

A key influence on voting behavior is peoples' family and friends. If all my family are lifelong Democrats, then I am a probably a lifelong Democrat too (same for Republicans). However, if my family all have different political allegiances, then that suggests that I might be more open to different perspectives and opinions, and therefore more likely to be a swing voter. So how could a politician make use of this type of information?

One thing they could do is to create a measure of the overall "Leftness" or "Rightness" of my family network. They assign people in my network a rating

of +1 if there is evidence that they are a Democrat and –1 if there is evidence of Republican tendencies. An average rating across my family network is then calculated. An average rating close to +1 or –1 indicates a strong family preference for Democrat or Republican respectively. A figure near to zero indicates my family are a mixed bunch: some Democrats and others Republican. If this is the case, then I'm probably open to persuasion one way or another and therefore the sort of person that it's worth spending time trying to convince to support a particular party. Consequently, if a political party can only afford to visit say, 10,000 people in a voting region containing 250,000 voters, then they will visit the 10,000 people whose scores are closest to zero.

Marketing is another area where social network analysis is used. In particular, to target key individuals within a network. A core concept in network analysis is centrality. It is possible to examine the connections between different entities and use this to identify the individual that is the "most connected," i.e. is most central to the network. A communication targeted at someone at the center of a network will spread out across the network far more effectively than if the communication is delivered to someone at the edge. Let's use my extended family network[4] as an example, as shown in Figure 9.1

As you can see from Figure 9.1, in my family my mother-in-law, Angela, is the person at the center, connected directly with almost everyone. That makes sense to me, because she is the one who acts as the go-between, talks to everyone all the time, arranges family get-togethers and so on. This is in stark contrast to myself, at the edge of the family network (as are Grant, Cathy and Mark). I rarely talk to anyone except my immediate family, and my mother-in-law is the one who lets my wife and I know what Bob, Grant and others (who I'm not directly connect to) are up to and vice versa. If a marketing company wanted to promote a new brand of chocolate (or whatever), then a good strategy might be to send a free sample to my mother-in-law, who, if she liked it, would be far more likely to tell everyone about it than I would.

A somewhat similar (and controversial[5]) application which builds on this idea of centrality is marketing products to children. Marketing organizations try to identify those with large networks of friends, and who are seen as trendsetters and influencers amongst their peers. Companies then send them free products: t-shirts, caps, posters and the like. The idea is that if they are seen to endorse a particular product, it will encourage the rest of their peer group to adopt that product too.

The examples presented so far demonstrate two different ways in which social network data is used. The political example was focused on the

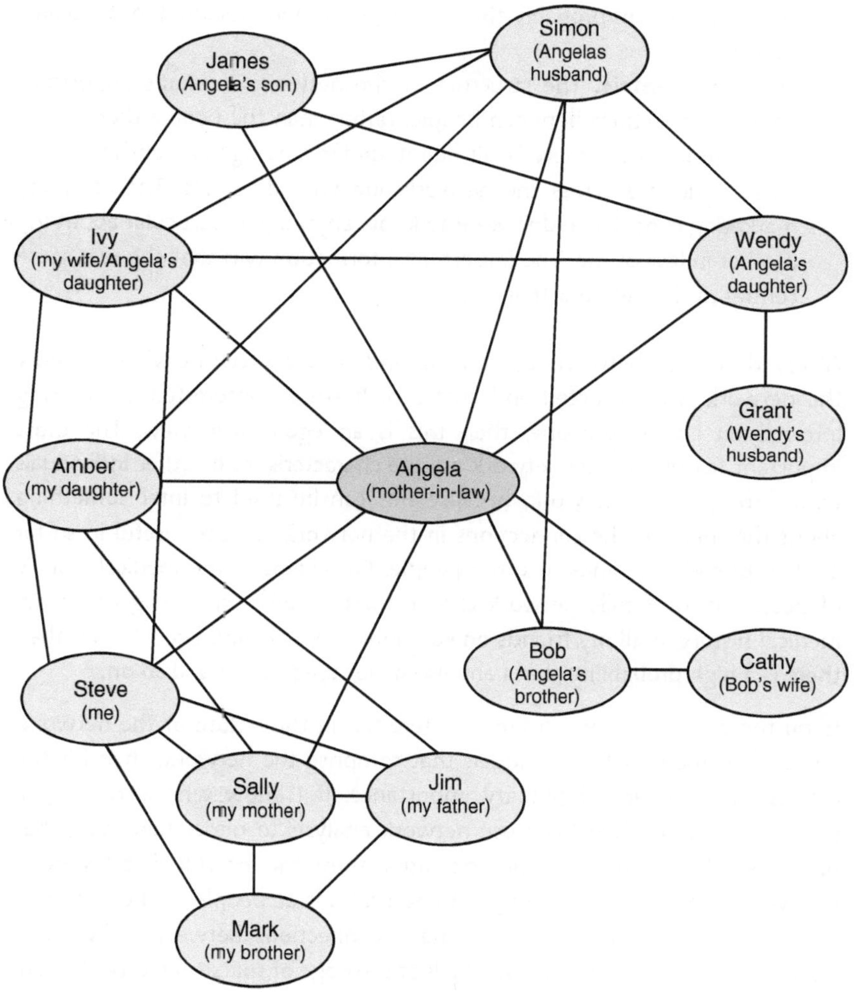

FIGURE 9.1 / Family network

network around me, whereas the marketing examples were more about how the network could be used to promote products more effectively across the network. More generally, we can categorize these two perspectives as:

1. **Ego-centric**. The focus is on the individual. The purpose of understanding a network is to learn more about the individual. Information about the network informs your view of someone, what they are like and what they do. If all my family are die hard Republicans, then I'm probably a staunch Republican too. If all my peers have a history of making lots of insurance

claims, then I am probably the sort of person who makes a lot of claims as well.

2. **Network centric.** The structure of the network is what's important, i.e. the connections between people, rather than the people themselves. A network centric view is all about understanding relationships: who exerts influence across the network and who does not. The company marketing chocolate didn't want to know anything in particular about my mother-in-law or me. The important information was that she was more central to the network than I was.

A key difference between ego-centric and network centric views is how the network is constructed and analyzed. If we are interested in inferring information about someone, then this is an ego-centric view. The most important features of the network are the characteristics of other individuals within the person's network, because this can be used to infer something about that person. The connections in the network are only useful in so far as they provide the links to other people. For example, the medical history of people in my family network can be used to infer something about my medical history. If all my friends on Facebook are male and aged 25–34, then there is a high probability that I am also male, aged 25–34 and so on.

If, on the other hand, we are more interested in the nature of the network, rather than the specific individuals that comprise the network, then it's the connections that are of primary importance. If I know who is carrying a particular disease, then I can use network analysis to understand how that disease will spread through the population from one individual to the next. Likewise, I can target marketing communications at people in the center of dense networks (where there are many connections between individuals) rather than wasting time on individuals at the edge of sparse networks (where each person is only connected to one or two others). Another application is using ideas such as centrality to identify those at the center of crime/ terrorist networks and target them, rather than picking off the small fry at the periphery.

In a predictive analytics context, ego-centric and network centric network views both have their uses, but it is the ego-centric view that is usually most important: it is information about someone's associates (their geo-demographics and primary and secondary behaviors) which provides the greatest value when developing predictive models. The connections in the network are the means by which the associates are identified. Information such as the number of connections someone has, whether someone is in a large or small network, in the center or at the edge and so on, usually adds far less to predictive models of individual behavior than associate information. However, there are some exceptions to this rule. For example, the connections are often very important predictors in fraud/crime models. The fact that an individual's credit card account is linked to 50 other card accounts, because they all have the same phone number, is a very strong predictor of fraudulent behavior.

9.7 Mapping a social network

Before you can do any social network analysis you need to establish who is in the network and how the people in that network are connected, i.e. you need to map the network. Unlike the connections in a microprocessor or a road network, it's not always easy to identify who is linked to who, how they are related or the strength of that connection. Creating a network view of a large population also requires a lot of computing power, particularly if one is taking an ego-centric view of individuals and their networks. This is because an ego-centric view requires that you examine the network that surrounds each and every individual, rather than building a network across a population, and then examining the properties of that network.

One reason why social network sites such as Facebook and LinkedIn have attracted so much attention, particularly from marketing organizations, is because they provide a ready-made network that the person themselves has created and shared – it provides a pretty good view of their family and friends, peers and other associates. Other types of network are more diffuse and difficult to map. In practice, what you usually end up with is a partial or incomplete view of a network based on the limited number of connections that you have been able to find, some of which may be more dubious/ uncertain than others. This doesn't invalidate the network; it just means that you might not be able to infer quite as much from it as you could if you had a more accurate and complete view.

So how do organizations identify connections between people? Social network sites aside, the most common way to identify connections is via "shared interests" and "common bonds." Some examples of shared interests/common bonds are:

- **Phone number(s).** These include the individual's number(s) plus numbers that they have recorded on mobile devices or elsewhere.

- **Addresses.** Current and previous residence, business addresses etc.
- **Family name**. This includes current surname, maiden name and other previous names.
- **Shared ownership.** For example, joint ownership of a property or having a joint bank account.
- **Shared responsibility**. This includes: being directors of the same company, signatories on an official document (e.g. insurance policy or marriage certificate), parent of the same child, convicted for the same crime and so on.
- **Transaction.** When someone gives or receives, buys or sells something from someone, it creates a connection between them.
- **Affiliation.** People who are members of a club, society or other organization, and in particular, if they are involved in the management or governance of that organization (i.e. shared responsibility).

So going back to Figure 9.1, my connection to my wife can be ascertained by our shared surname, because we live at the same address, have a joint mortgage, have a daughter together, have each other's numbers in our phones and so on. Given that there are multiple items of information providing evidence of a connection between us, this is gives a pretty strong guarantee that my wife and I are indeed part of the same network and are closely associated. However, not all types of connection are 100% guaranteed to provide a genuine link between individuals. Surname is a useful link item, but does not on its own guarantee that two people are from the same family. Likewise, two individuals may live in the same building, but rent their apartments from different landlords. Therefore it's often prudent to consider the strength of the connections in a network in addition to the fact that a connection exists. My link to my mother-in-law, for example, is more tenuous than to my own parents because I don't share the same surname and have never lived at the same address as her. So, if I were to map my family network again using only the strongest connections between individuals, most of the links with my wife's side of the family would disappear from Figure 9.1.

A key issue when mapping a network is the number of degrees of separation between individuals. Everyone is connected to everyone else eventually, and you need to draw a line somewhere to say where the boundaries of a particular network lie. Otherwise, you end up with huge networks that are difficult to understand and use. A network that describes one degree of separation contains information about people who have a direct connection with the individual in question. A network that considers two degrees of separation also contains information about people who are indirectly connected via someone else, i.e. there is an intermediary between those two people. In Figure 9.1 there is one degree of separation between myself and my wife, my daughter, my parents, my brother Mark and Angela. However, I am only connected to my wife's brothers and sisters (James and Wendy) via Ivy or Angela, i.e. there are two degrees of separation between James and myself and Wendy and myself. There is also one link in the family network with three degrees of separation from myself: my most direct connection with Grant is via Angela and then Wendy.

From a predictive analytics perspective, what you tend to find is that the information about people with one degree of separation is by far the most important, and the greater the strength of that connection the more useful the information about the associated individual will be. Sometimes information from two degrees of separation adds a little value, but in my experience not much, and it's rare for more distant connections (three or more degrees of separation) to provide any real value: one's nearest and dearest provide much more useful information than distant relations, friends of friends and such like.

chapter 10

Hardware, Software and All that Jazz

All large consumer-facing organizations maintain a number of different systems that hold data about people. A standard setup will include:

- **Operational systems.** These hold information that is directly relevant to managing the relationship that an organization has with its customers. Examples of operational systems include: seat reservation systems, power dialers,[1] insurance quotation systems and customer contact systems used for direct marketing. Most operational systems provide only limited analytical and reporting capability; usually just enough to allow the operational status of the system to be assessed: for example, how many cases were processed that day, total value of transactions in the system or number of calls waiting to be answered.
- **Enterprise (corporate) data warehouse(s).** A data warehouse contains all of the information that an organization deems to be important. It is used to drive analytical and business reporting functions. Ideally, all of an organization's data will be stored in a single data warehouse.
- **Reporting and analytical systems.** These extract data from the data warehouse and/or operational systems. The data is then available to be analyzed. The analytical systems are typically integrated with specialist software that enables predictive analytics and other types of data analysis to be performed.

Data within operational systems is sacrosanct. It can only be changed or altered in a very controlled way, lest the operation of the system is compromised – imagine the mayhem that would result if a data scientist working for Visa or MasterCard managed to delete even a day's worth of transactions or brought the system grinding to a halt because they had a big analytical job running. A data scientist may be given permission to extract copies of certain data from an operational system in certain circumstances (read-only access), but would never be allowed to alter, amend or change the data in such a system under normal circumstances.

All key data from the operational systems which needs to be maintained long term will be transferred to the data warehouse where it is maintained for as long as deemed necessary – which may be years or even decades for some types of data (e.g. account records and health records). As well as updates from all of the organization's operational systems, the data warehouse will also receive external feeds from organizations such as credit reference agencies, database marketing companies and affiliate organizations with whom they share data.

Some data warehouses are updated in (near) real time, but many are still updated in batch, either overnight or at month end, depending on the system. The data warehouse is an organization's key data resource, and like operational systems there will be controls in place to ensure that data cannot be overwritten and that people don't do "silly" things like fill up all of the available disk space with junk or set jobs running that never finish.

The management and control structures around corporate data warehouses are necessary but extremely frustrating to data scientists, whose job is to explore and experiment with the data. Consequently, standard practice is to have a separate analytical system that takes extracts of data from the data warehouse whenever analysis is required; if necessary, the data can be loaded on to standalone PCs or laptops for people to work with. A key feature of analytical systems is that they are removed from the day-to-day workings of the organization. Therefore data scientists can play and experiment to their heart's content without creating any operational risks.

In many organizations a key part of the analytical environment is the "sandpit." A sandpit is ideally a full copy of the organization's data warehouse, but in practice often contains only a sample of data in order to keep costs down. Data scientists can play to their heart's content in the sandpit to develop and refine their ideas and then, if necessary, run their code against the full data warehouse (in a controlled environment) to obtain the data and/or results that they need.

Traditional data warehouse solutions are good at moving data around, matching data and extracting it, and often have a lot of computer power to enable them to do this quickly with many users able to process jobs at the same time. However, they are not so good at providing analytical tools to understand the data or carryout predictive analytics. As discussed earlier, there is also the issue of protecting the data against accidental corruption or deletion, which means that data scientists can't operate as flexibly or efficiently as they would like within the data warehouse environment. On the other hand, analytical systems and sandpits are great for doing data analysis and building predictive models, but are often inferior in terms of their data processing and storage capabilities. In addition, there can often be bottlenecks and delays when data scientists want to transfer large quantities of data from the warehouse to the analytical environment.

A recent development[2] in data warehouse technology is in-database analytics. These systems seek to address these problems by combining the size and data processing capabilities of the warehouse with the flexibility of the analytical environment. These systems enable users to develop predictive models without needing to extract data to a separate analytical environment, but also provide the necessary safeguards to prevent the core data within the data warehouse being deleted or corrupted. Keeping all the data in the warehouse environment has a number of advantages:

1. **It supports good data governance.** Data is not distributed around the organization. This makes it less likely that data will be released unintentionally and means that everyone is working from the same base data. Often different versions of the data will sit with different people on different analytical servers/PCs, which can cause problems when it comes to reconciling results derived from different business areas.
2. **Speed.** There are potential savings in terms of the time it takes to complete analytical jobs. This is in part due to removing the requirement to extract data from the warehouse and in part to faster processing of the data within the warehouse environment.
3. **Scalability.** The warehouse is naturally a good place for Big Data analytics due to the scalability of the warehouse.

The main drawback of in-database analytics systems is that they don't yet offer the full range of functionality of specialist analytical software. However, vendors of in-database analytics continue to enhance their products, and analytical software providers such as SAS are adapting their software to work with in-database solutions. Consequently, the gap between in-database analytical systems and standalone analytical environments will no doubt continue to narrow.

10.1   Relational databases

There are many different types of database, but for many decades the business world has been dominated by relational databases. New technologies such as Hadoop are in the limelight, but most organizations still rely on relational databases to provide the backbone of their data strategies for managing customers. This is a situation that is likely to continue for the foreseeable future given the fundamentally different approaches to data management adopted by relational database management systems (RDBMS) and mass storage systems such as Hadoop.

Relational databases emerged in the early 1970s in response to inefficiencies in data storage at a time when even a few megabytes of storage cost thousands of dollars. This was in the days before smartphones and the Internet, and any customer correspondence would have been held in hardcopy format in rows of filing cabinets. There was no real concept of useful data that could not be stored in a "spreadsheet like" format of rows and columns. Consequently, most databases at that time maintained their data in a collection of unconnected tables – conceptually not dissimilar to a set of disparate spreadsheets. Each table was maintained independently of the others, which meant that data was often duplicated, creating a significant and unnecessary cost burden. A data table for a retailer of electrical goods, for example, might have looked something like the one shown in Figure 10.1

Figure 10.1 seems sensible enough – it's easy to see who bought what, when and so on – but there is a lot of duplicated information. A much more efficient way to store this data is to *normalize* the data and split it into two tables, as shown in Figure 10.2

In Figure 10.2, the Sales table from Figure 10.1 has been modified. The five fields containing product details (Manufacturer, product category, model, RRP and warranty period) have been removed and replaced with a single "Product ID" field. A second table has then been created containing product details. The Product ID field can be used to cross reference between the two tables, allowing the product details for any sale to be obtained.

Having two tables considerably reduces the amount of data held. For the purpose of this example, imagine that there are details of one million sales in the Sales table and the store stocks 20,000 different electrical products. In this case, the Sales table in Figure 10.1 would contain 5,000,000 cells of product data (1,000,000 sales × 5 pieces of product data). In Figure 10.2, there are only 1,120,000 cells containing product data.[3] Product ID is held for each purchase

Sales Table

Sale ID	Purchase date and time	Name	Date of birth	Manufacturer	Product category	Model	RRP	Warranty period
00000001	29/05/2013. 09.04	Scott Katerman	04/08/1978	Google	Tablet	Nexus 7	$199.99	12 months
00000002	29/05/2013. 09.59	Rosemary Lee	12/07/1960	Panasonic	Television	42GT60	$1,399.99	5 years
00000003	29/05/2013. 10.22	Steven Finlay	10/12/1974	Samsung	Blu-ray player	BD-F7500	$239.99	2 years
.....	.....	.....	.....	.....	.....	.....	.....	.....
00999998	25/04/2014. 17.20	Charlotte kid	31/10/1959	Panasonic	Television	42GT60	$1,399.99	5 years
00999999	25/04/2014. 17.28	Peter Harvey	21/11/1965	Dell	Laptop PC 17	Inspiron	$549.99	12 months
01000000	25/04/2014. 17.29	Ruth Burton	06/09/1987	Samsung	Blu-ray player	BD-F7500	$239.99	2 Years

FIGURE 10.1 Sales table

Sales Table				
Sale ID	Purchase date and time	Name	Date of birth	Product ID
00000001	29/05/2013. 09.04	Scott Katerman	04/08/1978	00000001
00000002	29/05/2013. 09.59	Rosemary Lee	12/07/1960	00019999
00000003	29/05/2013. 10.22	Steven Finlay	10/12/1974	00000002
......	......	......	......	......
00999998	25/04/2014. 17.20	Charlotte kid	31/10/1959	00019999
00999999	25/04/2014. 17.28	Peter Harvey	21/11/1965	00020000
01000000	25/04/2014. 17.29	Ruth Burton	06/09/1987	00000002

Many

1

Product Table					
Product ID	Manufacturer	Product category	Model	RRP	Warranty period
00000001	Google	Tablet	Nexus 7	$199.99	12 months
00000002	Samsung	Blu-ray player	BD-F7500	$239.99	2 years
......	......	......	......	......	....
00019999	Panasonic	Television	42GT60	$1,399.99	5 years
00020000	Dell	Laptop PC	Inspiron 17	$549.99	12 months

FIGURE 10.2 Many to one relationship

in the Sales table, and in the Product table there are five fields for each of the 20,000 product types. Given that many sales are for the same product, there is said to be a "Many to one" relationship between the Sales table and the Product table. In this example, there are just two tables, but in real life a

relational database may contain many hundreds of different tables, all holding different types of data.

Storing data across multiple tables in this way enables it to be held and managed efficiently and avoids duplication. Duplication is something all database managers seek to avoid, and within a well-constructed database any given piece of information is only ever held in one place. As well as being inefficient in terms of storage space, what you also tend to find is that if an item of data is stored in several different tables, perhaps across several different databases, differences arise. This occurs because when people make amendments to the data or change the way it's processed, they usually only update the table of immediate interest to them.

To make use of information across tables there must be at least one field (or combination of fields) in each table that has a unique value for each observation. This is referred to as a match key (or primary key in database terminology). In Figure 10.2, a primary key for the Sales table is Sale ID because each sale has a different Sale ID. The primary key for the Product table is Product ID. The Product ID field in the Sales table is referred to as a "foreign key." This means it is not a primary key for the Sales table (it's not unique for every row), but is a primary key for another table. Therefore it is possible to match records in the two tables together.

The most popular (although some would say not the best) tool used to manipulate relational databases is the programming language SQL (Structured Query Language). SQL was designed specifically to manipulate data in this format, making it easy to match, extract and manipulate customer data across several different tables. Using SQL it is very easy to access specific data items in a relational database, and to do all sorts of complex processing very quickly. One problem with SQL is that it's not the best tool for dealing with data that isn't nice and structured (e.g. text) or for matching data that is fuzzy (similar, but not an exact match).[4] That's not to say that you can't do these things using SQL, but SQL is not optimized for these types of application.[5]

10.2   Hadoop

Hadoop was the first popular Big Data environment and has become the tool set of choice in the Big Data community. The original version of Hadoop was developed by Doug Cutting and Mike Cafarella in 2005. It grew out of research by Google into mechanisms for storing and processing data from millions of web pages in a timely and efficient manner that was easily scalable as the amount of information grew. Cutting named it Hadoop after his son's toy elephant.[6]

The first thing to say about Hadoop is that it is not a single piece of software. Just as Microsoft Office is made up of Excel, Word, PowerPoint, Outlook and so on, so the Hadoop family encompasses a number of tools. The second thing to say about Hadoop is that you should not think about it as a database in the traditional sense, where the data is nicely prepared and formatted, processed and indexed to make it fit the requirements of the relational model. Instead you should view it as a way of storing massive amounts of diverse data and processing it quickly.

The foundation software you can't do without is the Hadoop Database Core tools (The Hadoop Common) and the Hadoop Distributed Filing System (HDFS). The HDFS manages the storage of data across any number of commodity servers. A commodity server simply means that it's an off-the-shelf computer that typically costs anything from a few hundred to a few thousand dollars.[7] There is nothing special about the hardware. If you want to increase your storage and data processing capabilities you just buy a few more servers.

A key feature of the HDFS is that you shouldn't think about any particular piece of data being stored on a particular server. One of the jobs of the HDFS is to maintain multiple copies of the data on different servers so that if one server fails, the data is not lost and a replacement copy can be generated and placed on another server. This is a very different way of protecting against server failures and data loss compared to the standard RAID[8]/regular backup model used with most relational databases.

The second core Element of Hadoop is MapReduce. MapReduce was developed by Google in 2004 as a way to split very large processing jobs into a number of smaller bits. Each bit is then processed by a separate server and the results combined together to deliver the final result. Versions of MapReduce can be used with other software, but it has become synonymous as the primary tool for interrogating Hadoop databases.

Hadoop has attracted so much attention, and has become very popular, very quickly, for the following reasons:

- **The software is open source.** You don't need to go out and haggle with vendors. If you want a copy to play with then you can download it onto your PC for free in a few minutes. Having said this, there are all sorts of specialist tools that vendors are now selling to enhance the Hadoop experience.

- **Hadoop runs on commodity servers.** Just like the Hadoop software, you aren't tied to one vendor's hardware. Any standard PC/server will do.
- **It's fully scalable.** Some organizations are running Hadoop across thousands of servers. If you need more storage you simply buy more, cheap, commodity servers.
- **It's very fast (for some things).** For some types of data processing, such as sorting large files, Hadoop runs amazingly quickly. If you have enough servers, some tasks can be processed 10–100 times faster compared to a traditional RDBMS installation.
- **It's designed to deal with structured and unstructured data.** Traditional RDBMSs are very efficient when it comes to processing structured data in rows and columns, but don't deal well with things like web pages and customer correspondence. That's not to say these things can't be managed within a RDBMS, but most RDBMSs are not very efficient at processing data in this format.

When it comes to Big Data Hadoop currently rules the roost, but there are alternatives. Two well-regarded competitors are HPCC,[9] and Storm.[10]

10.3   The limitations of Hadoop

Make no mistake, Hadoop is great, but everyone agrees that it's better at some things than others. Hadoop works best in write once, read many environments, i.e. once you've got data, Hadoop is a great at running analytical jobs across large swathes of it, to do things like sorting, counting and summarizing the data.

What Hadoop is less good at is real-time data processing on lots of small sub-sets of the data. Likewise, if you need to update customer records in real time, or need to retrieve specific data items quickly, then Hadoop is not the best tool for the job. This is because the greatest strength of Hadoop is also its greatest weakness. Being spread across many different servers, possibly on different sites at different geographical locations, one encounters the problem of latency. Programming jobs run very quickly once they get going, but there is a time overhead involved in splitting up the problem into chunks, distributing those chunks across all the different servers, and then pulling all the results together. Even if you have a job that requires only a fraction of a second of processor time, you may be waiting several seconds (or even minutes) for Hadoop to deliver the results. This is a significant issue for real-time data processing environments, such as call centers, transaction processing systems or web page interaction.

To improve the performance of Hadoop for certain tasks, products such as Hadoop Hive and Hbase have been developed. Hive mimics the features of traditional relational database management systems, using an "SQL like" language to manipulate data, and is designed to reduce the latency problem. However, the performance of Hive still lags behind a traditional RDBMS when it comes to real time customer interaction. This is because it still relies on the core HDFS, with data distributed across many different servers, making it better suited to large-scale batch processing.[11]

HBase, on the other hand, is designed to deal with data stored in massive tables and provides real-time read/write capability. HBase utilizes features of columnar databases to provide rapid performance, across columns of data, which for some tasks is a much more efficient mechanism than processing data in rows, as is common in relational databases. The downside of Hbase is that it lacks many of the data manipulation tools of relational databases. In particular, there is no concept of key fields within tables to allow matching across tables. Therefore it works very well if you can get all your data into one big table, but is less useful if data is spread across many different tables which you need to cross reference.

10.4   Do I need a Big Data solution to do predictive analytics?

The short answer is no. Big data solutions such as Hadoop, HPCC and Storm are pretty cool and I'm sure every data scientist wants their own implementation to play with, but the truth is that most business problems don't require Big Data solutions. There are far more "small" problems out there than big ones.

The majority of data analysis jobs (including predictive modeling) can be processed perfectly well using a traditional relational database linked to an analytical server. Even if you have petabytes (thousands of terabytes) of data, it does not mean you need to be processing all of that data all of the time. Even organizations with large and fast-moving databases, such as Amazon and Facebook, are typically only looking at relatively small sub-sets of what's available at any one time (a few gigabytes), in order to answer specific business questions. More often than not, they are doing this using laptops and desktop PCs.[12]

If the analysis database used to construct a model contains only a few gigabytes of mainly structured data, then there is not a huge business case for enhancing a traditional relational database solution with something massively parallel in order to do predictive analytics. It's only when your analytical databases get into the terabyte range and you have a lot of unstructured data that solutions like Hadoop come into their own.

To illustrate just how much data this is, let's imagine I've been asked to build some models for the largest supermarket in the UK. The objective is to predict which items customers are likely to purchase the next time they visit the store and when their next visit is likely to be. The idea is to e-mail and/or text them with tempting offers 1–2 hours before we think they will visit. If we have real time GPS information available then we can contact them as they arrive at the store, and if the GPS data tells us that they are *en route* to a rival supermarket, we may even text them with a fantastic discount offer to make them change their mind. A database containing the following information is put at my disposal:

1. Details of every purchase made by every customer at the supermarket in the last year (ten million customers buying an average of 50 items a week). This includes which store was visited, the date and time of each shopping trip, what items were purchased at what price and how much each customer spent.
2. 1,000 lifestyle variables about each of the supermarket's customers. So this is everything from their age and income, through to whether they are vegetarian or vegan, and where they like to go on vacation.

This equates to about 150 GB of data.[13] Make no mistake: this is a pretty big dataset. It contains lots of predictor variables that can be used to build some excellent models, but it is a dataset that can be stored, manipulated and analyzed using a single desktop PC. If processing time is a problem, then very little will be lost by using a 1:10, or even a 1:100, sample to do the analysis. However, if I now decide I want to consider ten years' worth of customer purchase data instead of one, and I want to start looking at all the websites that those ten million customers have visited in the last few months, plus information that I've obtained from loyalty card associates (other stores that share data with the supermarket) then that's the time things move into the terabyte range and beyond. If I want to use all of this data in its entirety, then a traditional relational database configuration is going to struggle, and it makes sense to consider a massively parallel database/analytics architecture such as Hadoop – at least for some of the heavy lifting during early stages of model development that involve collation and cleaning of the data.

I've mentioned it a number of times already, but one thing to remember is that when it comes to predictive analytics, most of the Big Data out there is fairly useless. If there are large swathes of your data landscape that have never been explored before, then it is well worth doing an audit of everything you have, and that will take a lot of resources the first time around. However, for almost all applications of predictive analytics there are two golden rules:

1. You can always establish what the important predictor data is using samples. My own rule of thumb is that no matter how large the dataset, if I can't determine whether a predictor variable is significant by looking at 100,000 cases, then it can't be that important.[14]
2. Once you've established which predictor variables are the important ones for predicting behavior, these tend to remain the important ones for a long time. You don't need to reassess each and every data item again – at least not very frequently.

There are always going to be some exceptions to these rules, but there aren't many. In Chapter 4 I said it's rare for a predictive model to contain more than 20–30 variables. What is also true is that for most applications of predictive analytics, and particularly those that deal with predicting consumer behavior, is that once you've found the important variables these continue to be the ones that are most predictive of that behavior.

The weights allocated to each variable will change over time and occasionally a new piece of data becomes available that can improve the quality of your models, and/or one data item may be substituted for another. However, even if data velocity is high, the relationships between data items change far less frequently. Blood pressure, age, weight and cholesterol have always been predictive of heart attack and probably always will be – regardless of how frequently you monitor the data or what other data becomes available. Your default history on past credit agreements is always going to be the most important predictor of future repayment behavior, previous claim history will always feature in an insurance claim model, when you last bought a product will always be an important predictor of when you next buy that product, and so on.

Having said this, it's important to appreciate that it's not a case of one or the other when comparing relational databases to something like Hadoop. Implementing Hadoop doesn't mean that you must throw away your existing databases – anything but. There is a growing trend towards hybrid solutions, utilizing relational databases for the most important and immediate data (such as maintaining customer account records and implementing predictive

models) and Hadoop for mass storage of more diffuse and unstructured data sources, to be analyzed and used when required.

One thing you can do is use Hadoop tools to do the preliminary leg work, to identify any important data amongst all the chaff and then shift the small proportion that is actually useful to your existing relational databases and operational systems to be used in real-time decision making. Off course what that means is that while Hadoop is relatively cheap, it's not going to reduce your IT costs because it's an add-on, not a replacement.

Suppliers of relational databases are also adapting their offerings to make them better able to deal with large unstructured data sources. Some are approaching this by creating better interfaces with Hadoop, while an alternative approach is to integrate Hadoop like functionality into existing database products and providing enhanced versions of SQL. Consequently, there is increasing overlap between the old and the new when it comes to storing, processing and analyzing data. At some point in the not-too-distant future it may no longer make sense to differentiate between the two. Hadoop is also massively scalable, and you can start small. So if you want to be ready to move into a Big Data world there is no reason why you can't start running Hadoop on a single server and then scale up as and when you need to.

10.5   Software for predictive analytics

There are lots of software packages that can be used to develop predictive models. Some of the most popular packages are SAS, R, IBM SPSS, Stata and RapidMiner. Most of these were originally designed to work on an analytical server or standalone PC/laptop once data had been collated from various sources. However, the vendors of these solutions now provide their software with the functionality to integrate with Big Data solutions such as Hadoop, and/or provide in-database analytical capability. There are also a number of new and emerging tools aimed specifically at Big Data analytics, such as Apache Mahout and Revolution R.

Most analytical software can apply a range of analysis and modeling techniques and is generally quite flexible, but each has its own strengths and weaknesses. Consequently, there are a number of factors to consider when deciding what software to use to do predictive analytics:

- **Data input/output.** How easy is it to get data into and out of the software? Can you, for example, load data directly from your organization's data warehouse, construct a model using the software and then export the results to another package or a spreadsheet?

• **Capacity limitations.** Does the software have any restrictions on the size or number of the data sets that can be processed? How many observations (rows) and variables (columns) can it handle? Some software is limited to the physical memory (RAM) of the server. Other software is more flexible and allows datasets to be spread across several hard drives and servers, and/or integrated with distributed architectures such as Hadoop.

• **Data Formats.** What data types can the software deal with? How well does it deal with unstructured data such as text, sound and images?

• **Data manipulation and variable creation**. Good software provides comprehensive and flexible tools to clean, sample and format data prior to modeling. It should also be possible to derive new variables (as discussed in Chapter 8), exclude observations that are not required and split the data into different sub-sets for model construction and reporting. Are tools for text mining and (social) network analysis provided?

• **Data transformation (Pre-processing).** How does the software format data prior to it being used to create a predictive model? A good package provides binning tools that automatically apply indicator (dummy) variables, weight of evidence or other transformations. For binning approaches, one should be able to manually adjust the binning and/or join together non-contiguous bins for business reasons.[15] Check that the software allows this.

• **Data visualization.** Does the software allow you to explore the data dynamically, using graphs and tables, to gain an understanding of the relationships that exist? Can these graphs be transferred to other packages such as Microsoft Excel or Word for inclusion within reports? Does the software allow you to produce reports in bulk? For example, if you have 1,000 predictor variables, can you produce analysis of all the variables in one go, sort them in priority or alphabetical order, and then export them into a report for people to refer to?

• **Variable reduction.** The software should provide tools to evaluate which variables are important, and which add no value and can be discarded prior to model construction. Good software allows the user to adjust the list of selected/rejected variables to ensure that certain ones are excluded (even if they appear to be predictive) and to force variables to feature in the model (even if there is little statistical evidence of their predictive ability).[16]

• **What types of model can be built?** Can many different types of model be constructed, such as linear models, decision trees, neural networks, clustering and support vector machines, or are you limited to one or two specific methods?

- **Scoring.** A common requirement is to apply the model to a new set of data that was not used to construct or validate the model. How easy is it to do this?
- **Model implementation.** After constructing a model, how will it be implemented? Some packages produce code in Java, SQL, C++ and so on, allowing the models to be inserted directly into the relevant operational system. A fast-growing trend is the use of Predictive Modeling Markup Language (PMML). This allows a model built using one package to be automatically implemented by another. PMML is currently supported by over 20 leading software vendors.[17]
- **Reporting.** Does the software facilitate the production of documentation required to satisfy any audit and/or regulatory requirements? Can reporting be provided for multiple holdout/out-of-time samples as well as the development and validation samples?

Cost is also something to bear in mind. Some modeling packages cost tens of thousands of dollars per annum in license fees. Others, such as R and RapidMiner, are open source and free, and my experience is that those that cost the most are not necessarily the best. R, in particular, has become the package of choice for many, with lots of add-on modules, supplied by a host of software vendors, providing all sorts of general and specialist analytical functions.

Another recommendation is always to look beyond the visuals of the package. Just because a piece of software has a nice graphical interface does not necessarily mean it's any better than a product with a more traditional code-based interface. If you want to do complex cutting edge analytics, then flexibility is key. One question that you should ask vendors is: does the software allow sufficient customization and data scientist interaction? Many packages provide highly automated features, allowing models to be constructed very quickly, but this can limit your options.

It is still the case that code-based model development software provides the most extensive and flexible toolsets, allowing data scientists to do far more than that offered by even the most advanced point and click interface. The downside is that considerable time is required to learn the coding language, and model development may take longer. A lot of modern software combines the best aspects of each approach. There is a nice user interface that people can start using with very little training, which is supported by a more in-depth coding side to the software for use by more experienced users.

Appendix A. Glossary of Terms

Accuracy. One measure of how well a model predicts. For classification, model accuracy is calculated as $(A + B)/(C + D)$. A is the number of times the model correctly predicted an event, B the number of times the model correctly predicted a non-event. C and D are the total number of events and non-events respectively. Accuracy is only one measure of model performance. There are many others including GINI (AUC), Somers' D, R-squared, Adjusted R-squared, Likelihood, Aitkin's Information Criteria (AIC), Maximum Absolute Percentage Error (MAPE), Absolute Percentage Error (APE), Average Error and Root Mean Squared Error (RMSE), to name but a few. (Note that many of these measures are also used to assess the performance of regression models, in particular R-squared, Adjusted R-squared, MAPE, APE and RMSE.)

Algorithm. A set of instructions, executed one after another, to solve a particular problem. In predictive analytics many different algorithms are used to derive the parameters of predictive models. Most of these algorithms are iterative. The algorithm starts with an initial set of model parameters (chosen by the user or assigned randomly), which are then modified by the algorithm at each iteration until a final solution is arrived at. Back-propagation, for example, is an iterative algorithm that can be used to derive the parameters of a neural network. However, not all predictive models are created using algorithms. For example, least squares, which is one of the most popular methods for finding the parameters of a linear model, is based on calculus (simultaneous equations).

Balancing. Many real-world data sets are imbalanced. This means that they contain far more events than non-events, or vice versa. Balancing is the process of modifying an imbalanced data set so that the numbers of events and non-events are equal. Certain algorithms used to generate predictive models perform much better when data sets are balanced, for example popular decision tree algorithms such as CHAID and C4.5. Others, such as logistic regression, are less sensitive to imbalance.

Big Data. Any data set that is too large and/or complex to be easily processed using a single PC/server; typically, at least a terabyte in size, and comprising a mixture of data types from a range of internal and external data sources.

Binning. The process of grouping cases into a number of groups or ranges. The properties of each bin are then used for analytical purposes, rather than the raw data. For example, a variable such as income can take many different values. For analytical purposes incomes are grouped into 20 ranges, each containing 5% of the population. Additional bins may also be created for missing cases, outliers or natural breaks in the data, such as at zero and 100% for ratio variables.

Categorical imperative. Immanuel Kant defined the categorical imperative as: "Do unto others as you would have done unto you."

Causation. The reason why something occurred. Not to be confused with correlation.

Class imbalance. *See* Balancing.

Classification and regression tree. A type of predictive model, created by recursively segmenting a population into smaller and smaller groups. Popular decision tree algorithms include CHAID, CART, and C4.5 and C5.0.

Classification model. A classification model generates a prediction as to whether a given event is likely to occur or not. For example, will someone make an insurance claim in the next 12 months, will someone die from a heart attack in the next five years or if you are likely to buy a particular book. The score from a classification model usually represents the probability of the event occurring.

Cluster. A set of cases that are grouped together on the basis of having similar characteristics.

Correlation. Two variables are said to be correlated if a change in one occurs in tandem to a change in another. This does not mean that one thing causes the other. Not to be confused with causation.

Credit scoring. Credit scoring was the first commercial application of predictive analytics and remains one of the most popular. It is used to assess the likelihood of an individual repaying what they borrow, i.e. their creditworthiness. In the credit industry distinctions are made between application scoring used to assess new customers applying for a credit product, behavioral scoring used to reassess existing customers for additional credit (credit line increase), and collections scoring, to assess the likelihood that a customer in arrears will repay what they owe.

Cut-off score. The decision boundary at which different decisions are made. Those scoring above the cut-off are treated one way, those scoring less than the cut-off are treated another way. In direct marketing, for example, people are only contacted about a product or service if their score exceeds the defined cut-off. Those scoring below the cut-off are not targeted.

Database. An organized collection of data. There are various different types of databases and different schema as to how the data is organized. Relational databases are the most popular type of database in everyday use, with data held in a number of separate tables (relations). Data in the tables is linked together using match keys such as account numbers, customer identifiers or other pieces of information that can be used to uniquely identify records within each table.

Data cleaning. The process of correcting, removing and codifying data that is incorrect, missing or incomplete. For example, removing test records from a dataset, coding missing values to default values or applying imputation algorithms to make an educated guess at what a value might have been where that item is missing.

Data mining. The analysis of large and complex datasets. The goal of data mining is to find interesting patterns in data that a human would be unable to identify easily from visual inspection. Data mining encompasses a wide range of automated procedures from statistics, computing and artificial intelligence/machine learning.

Dataset (dataset). A collection of data. The term is usually used to mean data that has been collated from one or more separate sources and held in a single file. For example, data extracted from several different tables in a relational database. The dataset is then used to perform various types of analysis. For example, the production of management information or the creation of predictive models. The major difference between a database and a dataset is that a dataset is usually a single file/table and is intended for a single specific purpose, whereas a database is a more strategic collection of data that may be distributed across many different tables/files.

Data scientist. A rebranding of a job role that for many years has been undertaken by data analysts, applied statisticians, data miners, operational researchers and management scientists, amongst others. A data scientist is someone who can combine knowledge of data, IT and analytics to deliver value-added business solutions. Good data scientists are pragmatic and practical in their outlook – focused on delivering useful solutions that work, rather than worrying too much about theory.

Decision tree. *See* Classification and regression tree.

Dependent variable. This is a representation of the behavior or thing that one is trying to predict, using a set of predictor variables. For classification problems the dependent variable must be a binary indicator, i.e. 1 or 0, yes or no. For regression problems the dependent variable is a numeric value that can take a range of values.

Decision engine. A software tool for implementing predictive models and decision rules.

Development sample. The set of cases (a dataset) used to construct a predictive model. Usually there will be at least two (and ideally three or four) different samples used when a model is constructed. Predictive analytics is applied to the development sample to determine the structure of the model. The model will then be tested on a separate validation (test) sample to obtain an unbiased estimate of how well the model performs. Ideally, the model will also be tested on additional holdout and out-of-time samples.

Discretization. *See* Binning.

Dummy variable. *See* Indicator variable.

ELT. An acronym for Extract Load Transform. For many analytical tasks a subset of data is extracted from an organization's data warehouse and loaded into a separate analytical server. The transform step is required to put the data into a suitable format for processing by the analytical software. Note that for some analytical tasks the transform process is performed before the load, in which case the acronym becomes ETL.

Ensemble. A collection of predictive models that all predict the same thing. A final assessment (prediction) is made by combining the predictions from the individual models.

Expert system. A rule-based prediction system. An expert system comprises a rule base (knowledge base), a user interface and an inference engine. The user interface gathers information from the user, which is then used by the inference engine to see which rules fire in order to come to a conclusion/make a prediction.

Forecast horizon. Most (but not all) predictive models forecast future unknown events or quantities. The forecast horizon is the time period over which a model predicts. For example, insurance claim models typically forecast claim behavior over a 12 month forecast horizon. The forecast horizon for a mortgage default model is typically 18 or 24 months.

Gain chart. A popular tool for demonstrating the benefits of a model. See Appendix C for further details.

Gigabyte. A gigabyte is 1,000 megabytes (see below). A gigabyte of storage can hold around one billion characters of text. This is enough for more than 1,000 books, several hundred pictures or about an hour of standard definition film.

Hadoop. A family of open source tools for storing and processing Big Data using cheap, off-the-shelf "commodity" computers. Hadoop is highly scalable, fault tolerant and relatively cheap compared to traditional "super computer" approaches to processing huge amounts of data.

Holdout sample. An independent sample of data that has not been used to construct a predictive model. The holdout sample is used to test the model and provide a representative picture of how the model performs. Not to be confused with validation sample.

HPCC. High performance computing cluster. An alternative to Hadoop, providing a mechanism for storing and analyzing large datasets which are distributed across many different PCs/servers.

In-database analytics. A system that fuses together the data storage and processing capabilities of a data warehouse, with analytical tools for analyzing and reporting on the data. This removes the need to extract the data to separate analytical servers or standalone workstations.

Indicator variable (dummy variable). A common approach to transforming categorical data, and complex non-linear data, into an easy to use linear form. For categorical variables, such as residential status, a separate indicator variable is created for each category (home owner, renter, living with parents etc.). For numeric data such as "Value of property" the variable is split into a number of ranges (bins), and a different indicator variable used to represent each range.

Latency. The wait time between a task being requested and it being carried out. Latency can be a problem for some Big Data solutions (such as Hadoop) because of the time required to split jobs into chunks, distribute the chunks across the network of servers and then wait for the results to be fed back.

Lexicon. A collection of words. Lexicons are commonly created to contain words that relate to a specific topic. For example, legal, mathematical and management lexicons contain only words about those disciplines. In sentiment analysis, lexicons are widely employed that contain lists of words commonly associated with "positive" and "negative" sentiments.

Lift chart. A popular tool for demonstrating the benefits of a model. See Appendix C for further details.

Linear model. A very popular type of predictive model that is easy to understand and use.

Linear regression. The most popular method for creating regression models. Linear regression is based on the principle of least squares. The model is chosen (using simultaneous equations) that minimizes the sum of the squared error; where the error is the difference between what the model predicts and what actually happened.

Logistic regression. The most popular method for creating classification models. Logistic regression is based on the principle of maximum likelihood.

Machine learning. Machine learning algorithms are based on research into artificial intelligence and pattern recognition in order to learn from data. Examples are the algorithms used to train neural networks and support vector machines. Machine learning is often considered to be a sub-set of data mining.

MapReduce. A programming paradigm for segmenting computational problems and using different computers to solve each segment of the problem. The segments are then brought back together again to deliver the complete solution. Sorting large lists of data is a classic problem that benefits from this type of segmented/parallel processing approach. MapReduce and similar software is often applied in conjunction with Big Data solutions such as Hadoop and HPCC.

Megabyte. A megabyte is a measure of digital storage capacity. One megabyte is sufficient to store about one million characters of text or a single photo.

Metadata. Information about data. Metadata describes what type of data a database holds, together with detailed information about each piece of data within the database – what it contains and how it is formatted.

Model. *See* Predictive model.

Modeling objective. *See* Dependent variable.

Neural network. A type of model that is well suited to capturing complex interactions and non-linearites in data.

Neuron. The key component of a neural network, analogous to a neuron in the human brain.

Odds. A way to represent the likelihood of an event occurring. The odds of an event is equal to $(1/p) - 1$ where p is the probability of the event. Likewise, the probability is equal to $1/(odds + 1)$. So, odds of 1:1 is the same as a probability of 0.5, odds of 2:1 a probability of 0.33, 3:1 a probability of 0.25 and so on.

Open source. Software is open source if the programming code is made freely available (often at no charge), allowing anyone to make changes to it. With popular open source software a standards board is usually set up to control updates to approved release versions of the software. The Apache Software Foundation is one of the leading organizations providing open source software solutions, including Hadoop.

Outcome period. *See* Forecast horizon.

Outlier. A value that lies outside the usual or expect range. There are no hard and fast rules for identifying outliers. However, in traditional statistics, one (simplistic) definition of an outlier is an observation that lies more than 2–3 standard deviations from the mean.

Out-of-time sample. This is a holdout sample, taken from a different time period to the model development sample. The performance of the model is tested on the out-of-time sample to see how the model performs over time. Usually (but not always) the out-of-time sample is taken from a period just before or just after the model development sample.

Over-fitting. This is when the predictive analytics process finds patterns that exist only in the development sample and which are not replicated in the wider population. Over-fitting can be identified by comparing model performance on the development sample against an independent holdout sample. If model performance is significantly worse on the holdout sample, then this is a strong indication that the model has been over-fitted. As a rule, over-fitting becomes more likely as the ratio of parameter coefficients (weights) in the model to the number of observations in the development sample increases. One rule of thumb for linear models is that over-fitting is very likely if the ratio is greater than 0.1. However, this is only a very rough guide, and significant over-fitting can occur in samples where the ratio is much lower than 0.1.

Override rule. A rule used to decide upon a course of action, regardless of the prediction made by the model. For example, always decline credit applications from people aged under 18, no matter how good their credit score.

Over-sampling. A method for creating a balanced data set by replicating cases of the minority class. Over-sampling generally yields better results than under-sampling, particularly when there are relatively few examples of the minority class.

Parameter coefficient. A weighting that is applied to a variable within a predictive model. The larger the parameter coefficient, the more that variable contributes to the model score.

Petabyte. 1,000 terabytes. The largest commercial databases in existence today (as at 2014) are around 30 petabytes.

Predictive analytics (PA). The process of determining important relationships between items of data to aid in the prediction of future (or otherwise unknown) outcomes. The relationships that are found by the predictive analytics process are captured in the form of a model. The model can then be used to make new predictions.

Predictive model. A predictive model captures the relationships between predictor data and behavior, and is the output from the predictive analytics process. Once a model has been created, it can be used to make new predictions about people (or other entities) whose behavior is unknown.

Predictive Model Mark Up Language (PMML). PMML is a set of standards about how predictive models are coded. This is to facilitate automatic transfer of models from the analytical system used to create them to the operational system where they are deployed.

Random forest. A very popular ensemble method. Random forests are usually constructed from a large number of small decision trees, with each decision tree constructed under a slightly different set of conditions. Although most commonly applied to decision trees, the same principles can be applied to most forms of predictive model.

Regression model. A popular tool for predicting the magnitude of something, such as how much someone will spend or how long they will live. This is in contrast to a classification model which predicts the likelihood of an event occurring: for example, if someone will survive another five years, or whether they buy a particular product or not.

Relational database. A popular way of storing data efficiently across many tables (relations). Unique match keys (primary keys) allow information to be matched across tables as and when required.

Response (choice) modeling. A response model predicts the likelihood that someone will respond to a prompt, typically a marketing communication trying to sell them a product or service. Response modeling and credit scoring are probably the two most widely used applications of predictive analytics in the world today.

Sampling. The process of creating a sub-set of the available data. Sampling is applied for several reasons. One reason is to create a manageable dataset that can be analyzed quickly and easily. Another is to create a dataset that is representative of the operational environment where the model is going to be deployed.

Sandpit. A standalone system or an isolated part of a data warehouse, where analysts can do whatever they want with data, without any risk to the operational environment or data held within the organization's data warehouse.

Score. A number generated by a predictive model that represents the behavior that the model predicts.

Scorecard. A scorecard is a popular type of (linear) model that is widely used in many industry sectors. A scorecard describes which predictor variables are important and how many points should be allocated to each attribute of a variable. A predictive score is calculated by simply adding up the points that apply.

Sentiment analysis. Analysis to determine people's attitude towards something: for example, if someone likes or dislikes a particular product, is a supporter of left or right political parties, if they had a positive or negative experience when talking to a customer services representative. Traditional sentiment analysis is carried out using questionnaires and surveys. In a Big Data/predictive analytics context, sentiment analysis involves extracting information from text or speech to create data items that can then be used to construct a predictive model.

SQL. Structured Query Language. The most popular programming standard for manipulating data within a relational database.

Storm. An open source Big Data alterative to Hadoop, promoted as offering superior real-time data processing capability.

Support vector machine. An advanced form of non-linear model that has some similarities with neural networks.

Target variable. *See* Dependent variable.

Telematics. The use of wireless technology to gather data about people's movements. One well-known application of telematics is in "black box" recorders, placed inside a vehicle to capture information about the speed, acceleration, braking patterns and so on. Insurers use telematic data to improve the accuracy of their claims models, and hence what premiums to charge. Telematics can also be used in conjunction with other mobile technology (e.g. smartphones) to track people's movements, independent of their means of transport.

Terabyte. A terabyte is 1,000 gigabytes. A terabyte of storage can hold around 1,000 billion characters of text. This is enough for more than one million books, several hundred thousand pictures or about 1,000 hours of standard definition film.

Under-sampling. A method for creating a balanced development sample from a highly imbalanced population, i.e. a population that contains far more events than non-events or vice versa. With under-sampling, a balanced sample is created by randomly removing some of the more frequently occurring cases. As a rule, under-sampling is usually inferior to over-sampling, particularly when there are relatively few cases of the minority class.

Utilitarianism. An ethical philosophy that puts forward the view that an ethical action is one that leads to an increase in the overall happiness of the population.

Validation (test) sample. An independent data sample that is "put to one side" when constructing predictive models, i.e. the data is not used to construct the model. The validation sample is then used to test how well the model performs. A validation sample is usually used during model development to test interim models before a final model is agreed upon. Once the model has been finalized, final validation checks and performance metrics are calculated using holdout and/or out-of-time samples.

Appendix B. Further Sources of Information

Non-technical/introductory resources

The following sources are suitable for the general reader. They are relatively jargon free and make very few assumptions about the reader's prior knowledge. There is also very little in the way of mathematics/statistics/formulas or technical gobbledygook.

Siegel, E. (2013). *Predictive analytics: The power to predict who will click, buy, lie, or die.* **Wiley.** This is not in any way a technical book, nor in my opinion does it provide much in the way of practical advice about how to develop and use predictive models. However, that's not what it's about. It gives a real flavor of the art of the possible in a lively and entertaining way. You can't help but feel the author's enthusiasm for the subject.

Hurwitz, J., Nugent, A., Halper, F. and Kaufman, M. (2013). *Big data for dummies.* **Wiley.** I found this to be one of the more accessible introductions to Big Data. It also provides simple descriptions of the Hadoop family of Big Data tools and how Hadoop can be applied to business problems.

Silver, N. (2012). *The signal and the noise: Why so many predictions fail.* **Penguin.** This is a great book about forecasting, which provides a lot of invaluable advice about the weaknesses in our understanding of prediction systems. Although not specifically about predictive analytics or Big Data, there is a lot of wisdom here that anyone dealing with predictive analytics can learn from.

Kahneman, D. (2012). *Thinking, fast and slow.* **Penguin.** This book provides a lot of insight into how people think when making decisions. It gets to the heart of some of the reasons why people find it difficult to accept decision-making tools such as predictive analytics, and discusses a host of other interesting questions about the thought processes people go through when deciding what to do.

Information commissioner's office (2012). *Anonymisation: Managing data protection risk code of practice.* **Information Commissioner's Office.** http://ico. org.uk/for_organisations/data_protection/topic_guides/~/media/documents/library/

Data_Protection/Practical_application/anonymisation-codev2.pdf. This is an online publication published by the UK data protection regulator. It describes a pragmatic approach to creating anonymized data sets in compliance with EU data protection law.

Franks, B. (2012). *Taming the big data tidal wave: Finding opportunities in huge data streams with advanced analytics*. Wiley and SAS Business Series. Wiley. This provides a gentle introduction to Big Data and Hadoop. There is a little bit of discussion around predictive analytics, but it's really a Big Data book that introduces the reader to some of the key ideas behind Big Data solutions such as cloud computing and massively parallel approaches to data. You can probably read this in a couple of hours, making it ideal for the train or plane.

Cukier, K. (2013). *Big Data. A revolution that will transform how we live, work and think*. John Murray. This is another easy-going and very readable introduction to the world of Big Data and the opportunities for leveraging it.

Chryssides, G. D. and Kaler, J. H. (1993). *An introduction to business ethics*. Chapman and Hall. This book may have been written before the Internet revolution, but for a book about ethics it's relatively easy to read and contains a good number of case studies. Despite its age it continues to provide a comprehensive overview of ethics and its application to business. Used copies are cheap to come by these days. My copy cost less than $5!

The middle ground

These books/websites have some specialist content – maybe some math and a few formulas, or some IT technical speak, but not an overwhelming amount. Even if you avoid the technical stuff, then there is probably still a good deal that you can get from these books.

Linoff, G. S. and Berry, M. J. (2011). *Data mining techniques: For marketing, sales, and customer relationship management*. 3rd Edition. Wiley. This is a broad, well-rounded, and not overtly technical book that describes the most popular data mining techniques applied to direct marketing.

Finlay, S. (2012). *Credit scoring, response modeling and insurance rating*. 2nd Edition. Palgrave Macmillan. This is one of my earlier books. It expands upon the themes discussed in Chapters 6–8, focusing on the process of building, implementing and using predictive models. The focus is very much on the use of predictive models in financial services and retailing. However, the methods and principles of model construction that it describes are widely applicable across very many industry sectors.

Witten, I. H., Frank, E. and Hall, M. A. (2011). *Data mining: Practical machine learning tools and techniques*, 3rd Edition (The Morgan Kaufmann series

in data management systems). This is a detailed reference manual for those interested in practical data mining. I found it provided a nice blend of theory and practice, with many good examples.

Scott, J. G. (2012). *Social network analysis.* **3rd Edition. Sage.** A relatively short and concise introduction to social network analysis.

Easley, D. and Kleinberg, J. (2010). *Networks, crowds, and markets: Reasoning about a highly connected world.* **Cambridge University Press.** Provides a blend of practical and theoretical material about the application of social network analysis in a number of environments.

Wasserman, S. and Faust, K. (1994). *Social network analysis: Methods and applications.* **Cambridge University Press.** Provides a good grounding of the basic principles of network analysis in the early chapters, with more detailed theoretical material featuring in the later part of the book.

White, T. (2012). *Hadoop: The definitive guide.* **3rd Edition. O'Reilly Media.** There are not many books that provide a solid introduction to Hadoop, but this is a reasonable place to start. The Hadoop environment is still evolving rapidly and this edition may be a little out of date by the time you read it, but I look forward to a new, fully updated edition in due course.

Analytic bridge. http://www.analyticbridge.com/. This is a great resource and discussion forum for people with questions about predictive analytics and data science. Anyone can become a member and post questions/answers, as and when they wish. Discussions range from the very theoretical/technical, to practical application and discussions around "mundane" but essential aspects of predictive analytics such as data cleaning.

LinkedIn. http://www.linkedin.com/. Another free resource that anyone can signup to for free. If you don't know about LinkedIn, the best way to describe it is as a "Facebook for professionals." There are several predictive analytics forums on LinkedIn that discuss topics relating to predictive analytics and Big Data.

Menard, S. (2002). *Applied logistic regression analysis.* **2nd Edition. Sage.** This provides a short practical guide to using logistic regression. There are many newer algorithms that are theoretically superior to logistic regression when it comes to constructing classification models. However, logistic regression remains the most popular method employed in practice.

Zikopoulos, P. C., Eaton, C., DeRoos, D., Deutsch, T. and Lapis, G. (2012). *Understanding big data. Analytics for enterprise class Hadoop and streaming data.* **McGraw-Hill.** This a relatively short e-book about Big Data that you can download for free from IBM. Given that it's written by IBM staff, there is inevitably a focus (bias?) on IBM's own Big Data solutions. However, I can forgive the fact that I counted more than 300 references to IBM in its 166 pages because it's not badly written, contains a lot of simple material that can be read in a couple of hours, and it's free. Just remember that there are other providers of analytical and Big Data solutions out there!

Meys, J. and de Vries, A. (2012). *R for Dummies.* **2nd Edition. John Wiley & Sons.** R has grown to be one of the most popular statistical programming languages. It's widely used in academia and increasingly in the private/public sector. It's open source and free, so if you want to do predictive analytics at home you can be up and running in a few minutes (http://www.r-project.org/). This is not as in depth as "The R Book" by Crawley (in the next section), and you need to look further afield to get fully up to speed with using R to do predictive analytics. However, it provides a good simple introduction to the R package and is aimed firmly at the beginner who wants to get to grips with the fundamentals. As these types of book go, it's not that expensive at around \$20/£15/€20.

Academic/technical resources

My recommendation is that you'd benefit from a good grounding in mathematics or statistics, prior knowledge of data mining and/or have some knowledge of IT systems and/or programming languages if you want to tackle the material in this section.

Hosmer, D. and Lemeshow, S. (2013). *Applied logistic regression (Wiley series in probability and statistics).* **3rd Edition. Wiley.** This book provides a detailed look at the theory and application of logistic regression, which remains the most widely applied method for generating classification models.

Bishop, C. M. (1995). *Neural networks for pattern recognition.* **Clarendon Press.** This is one of the few definitive guides to the theory and application of neural networks. Although it was originally published back in the 1990s, most of the material remains as relevant as it was when it was first published.

Hastie, T., Tibshirani, R. and Friedman, J. (2011) *The elements of statistical learning: Data mining, inference, and prediction,* **2nd Edition. Springer.** This is a heavy weight guide to many of the data mining tools used in predictive analytics, written by three world-leading academics.

Bishop, C. M. (2007). *Pattern recognition and machine learning (Information science and statistics).* **Springer.** This book covers a lot of the theoretical material underpinning many of the tools commonly used for data mining and predictive analytics.

Crawley, M. (2012). *The R book.* **2nd Edition. Wiley.** This book is comprehensive, but also suitable for relative beginners (with some rudimentary experience of programming languages – maybe Visual Basic, C++, SAS, Java or Python), as well as more experienced statistical programmers who wish to learn how to use the R programming language.

Edmond, R. and Wilson, J. (2012). *Seven databases in seven weeks: A guide to modern databases and the NoSQL movement.* **Pragmatic Bookshelf.** As the title says, this book provides a great introduction to seven types of database that are available as open source products (so no product tie-ins – unlike many other books). The emphasis is on some of the "new" database technologies that are proving to be of real benefits to organizations looking to manage their data in a different way to the traditional relational/SQL model. However, PostgreSQL is one of the seven database technologies that is covered (a popular open source relational database supporting SQL).

Appendix C. Lift Charts and Gain Charts

Two of the most popular graphical tools used to assess classification models are lift charts and gains charts. To illustrate these, let's start with an extended version of the score distribution table used in the Booles case study from Chapter 2, as shown in Table C.1.

The score distribution now includes descending cumulative columns – and for expediency only columns relating to the full contact list have been retained, i.e. the columns relating to the test campaign have been removed.

A lift chart is produced by plotting the proportion of the population scoring above the cut-off score, against the overall response rate for those cases above the cut-off (Columns H and I in Table C.1). This lift chart is shown in Figure C.1.

The solid line on the lift chart is the base rate: the response rate that one would expect from a completely random direct marketing campaign or (to put it another way) the response rate that would be expected from mailing the entire contact list. The dotted line shows the trade-off between the proportion of the contact list that is targeted, and the response rate that would be expected from targeting that proportion.

As you would expect, the smaller the proportion targeted, the higher the response rate. To put it another way, if you only target a few very good cases then you will get better response rates than targeting a larger population. If you want to target more cases, then you have to include worse groups, resulting in a lower overall response rate. One way that the lift chart is used is to compare different cut-off strategies such as: "*If we target the top 10% of the population, then the expected response rate will be 4.6%. However, if we target the top 30% of the population, then the expected response rate will be 3.3%*" and so on.

A gain chart is similar to a lift chart. It is produced by plotting the proportion of the population scoring above a given cut-off score against the proportion of the total responses scoring above that score (Columns H and K in Table C.1). The gain chart generated from Table C.1 is shown in Figure C.2.

Table C.1   Amended score distribution

| | | Score distribution | | | | | Descending cumulative score distribution | | | | |
A	B	C	D	E	F	G	H	I	J	K
Score (end node)	Number contacted	% Contacted	Response rate %	Expected number of responses	% Responses	Number contacted	% Contacted	Response rate	Expected number of responses	% Responses
1	780,850	16%	0.500%	3,904	5%	5,000,000	100%	1.600%	80,000	100%
2	840,995	17%	0.613%	5,159	6%	4,219,150	84%	1.804%	76,096	95%
3	685,088	14%	0.823%	5,639	7%	3,378,155	68%	2.100%	70,936	89%
4	690,006	14%	1.212%	8,364	10%	2,693,067	54%	2.425%	65,298	82%
5	556,930	11%	1.667%	9,282	12%	2,003,061	40%	2.842%	56,934	71%
6	424,977	8%	2.247%	9,550	12%	1,446,131	29%	3.295%	47,652	60%
7	322,605	6%	2.740%	8,838	11%	1,021,154	20%	3.731%	38,102	48%
8	199,100	4%	2.941%	5,856	7%	698,549	14%	4.189%	29,263	37%
9	236,807	5%	3.509%	8,309	10%	499,449	10%	4.687%	23,408	29%
10	182,602	4%	4.762%	8,695	11%	262,642	5%	5.749%	15,099	19%
11	80,040	2%	8.000%	6,403	8%	80,040	2%	8.000%	6,403	8%
Total	5,000,000	100%	1.600%	80,000	100%	5,000,000	100%	1.600%	80,000	100%

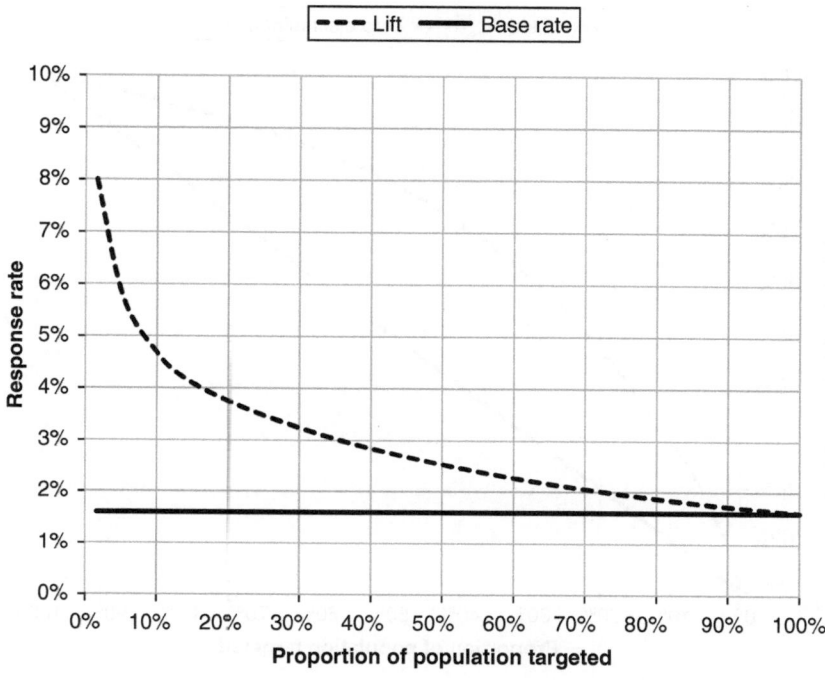

FIGURE C.1 A lift chart

The gain chart is useful for determining the proportion of events that will occur in the highest scoring *X*% of the population. For example, Figure C.2 can be used to determine that 30% of all responses can be achieved by targeting just 10% of the population.

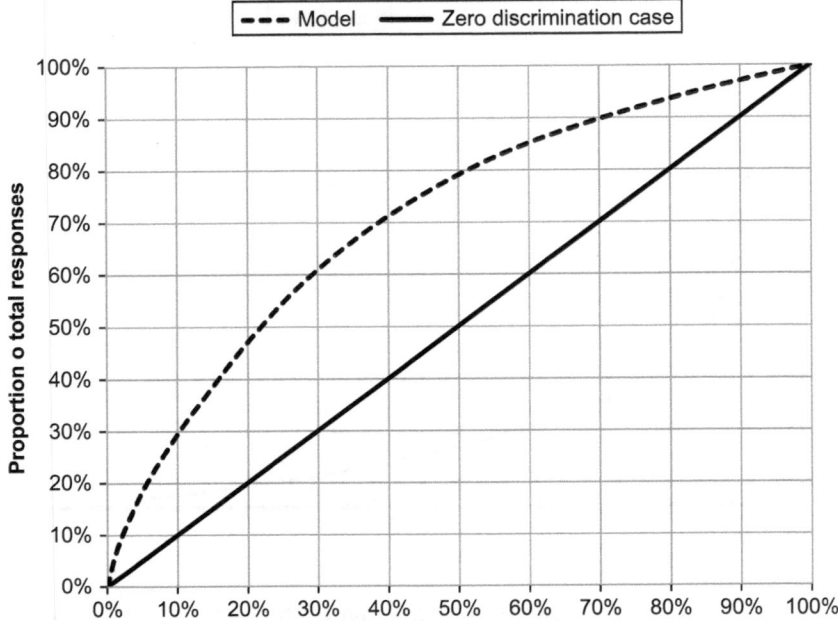

FIGURE C.2 / **A gain chart**

Notes

1 Introduction

1. The first credit scoring research was commissioned by the US government. Durand, David. (1941). *Risk Elements in Consumer Instalment Financing*. National Bureau of Economic Research. Durand used discriminant analysis (still a popular method today) to derive models to predict credit risk, using a sample of more than 7,000 historic loans. This was no mean feat given that the electronic computer hadn't even been invented in 1941.
2. The precise range depends on which FICO score is used. There are different versions for credit cards, loans, mortgages and so on.
3. A scorecard like this is created by doing some simple transformations (calibrations) on the output from standard statistical procedures for generating linear models. This includes, amongst others: logistic regression, Tobit and probit models, survival analysis, linear regression and least angle regression (LARS).
4. See *Thinking, Fast and Slow*, by Daniel Kahneman (2012), for some great examples of this type of behavior.
5. Eric Siegel lists 127 different applications of predictive analytics in his 2013 book, *Predictive Analytics: The Power to Predict Who Will Click, Buy, Lie or Die*, and even this is only a subset of the applications to which predictive analytics is being put.
6. Odds and probability are different ways of saying the same thing. To swap between the two: odds = (1/Probability) − 1. Probability = 1/(Odds + 1). For example, a probability of 0.25 is the same as odds of 3:1 and odds of 9:1 = a probability of 0.1.
7. In most countries regulators require models to be independently validated by someone other than the model developer. Regulators will also probe very deeply into how data was gathered, how the model was constructed and how it is going to be used. A very significant amount of effort therefore goes into documenting models and ensuring that they satisfy the regulator's requirements.
8. The first three of these properties of Big Data are often referred to as "The three Vs" and were originally formulated by Doug Laney in his 2001 paper "3D

Data Management: Controlling Data Volume, Velocity and Variety." http://
blogs.gartner.com/doug-laney/files/2012/01/ad949-3D-Data-Management-
Controlling-Data-Volume-Velocity-and-Variety.pdf. Accessed 24/03/2013.

9. A decent PC can store this amount of data easily. However, trying to manipulate
and analyze a terabyte of data is another story altogether, even using the best
PCs currently available (as at 2014).

10. The first three of these form the traditional definition of Big Data. Some
commentators also talk about the veracity (accuracy) of data, which is particularly relevant to externally sourced data where an organization has little control
over its form and content. IBM was one of the first organizations to talk about
the veracity of data as an issue for Big Data.

11. It's only when one starts to process data sets in the terabyte range that distributed systems such as Hadoop start to demonstrate superior performance to
traditional IT systems. Hadoop is also suited to certain types of data processing
tasks. It's not so good when it comes to rapid transactional processing, such as
maintaining credit card account records.

12. This is generally the case, but if the data is of poor quality then extra data may
not add much value over what you already have.

13. Another way in which Big Data is used is identifying data that is related in
some way. For example, the intelligence community is interesting in finding
all information that relates to specific threats to national security. They want
to look for common patterns across large numbers of documents, web pages,
transcripts and so on to identify those that relate to the threat. In a similar vein,
academics are interested in document search methods (text mining) to produce
lists of papers relating to specific research topics that produce better lists than
those generated from using simple keyword searches. Another application is
predicting stock market movements, based on sentiments being expressed in
blogs, tweets and so on.

14. I can't claim to be the first to use a gold analogy for data mining/Big Data. For
example, see: Zikopoulos, P. C., Eaton, C., DeRoos, D., Deutsch, T. and Lapis,
G. (2012). *Understanding Big Data. Analytics for Enterprise Class Hadoop and
Streaming Data*. McGraw-Hill.

15. Estimates about how much gold is in seawater vary. The most recent studies
now put it at a few thousand tons: Falkner, K. and Edmond, J. M. (1990). Gold
in seawater. *Earth and Planetary Science Letters*. 98(2). 208–221. Some previous
estimates put the figure at more than a million tons. The amount of gold mined
in all of human history up to 2011 is about 171,000 tons. http://en.wikipedia.
org/wiki/Gold. Accessed 17/01/2014.

16. This isn't a dig at Big Data evangelists. Whenever a new technology comes
along it tends to get over-hyped and the benefits overstated. As the technology becomes established people begin to realise the downsides as well as the
upsides and a more balanced picture emerges. The Gartner Hype cycle captures
this concept exceeding well. http://www.gartner.com/technology/research/
methodologies/hype-cycle.jsp. Accessed 17/01/2014.

17. This is the reduction in the error rate, not the absolute improvement in accuracy. For example, if you predictive models are 80% accurate, then this might increase to 82%. The error rate drops 10% from 20% to 18%.

18. The value of implementing a Big Data approach across government is estimated to be 2.5–4.5%: Yiu, C. (2012). The Big Data Opportunity. Making government faster, smarter and more personal. The Policy Exchange. http://www.policyexchange. org.uk/images/publications/the%20big%20data%20opportunity.pdf. Accessed 17/01/2014. Companies that combine Big Data with analytics, and who adopt a data-driven strategy, can expect to see increases in productivity/profitability of about 5–6%: MacAfee, A. and Brynjolfsson, E. (2012). Big Data: The Management Revolution. *Harvard Business Review*. October 2012.

19. Since 2011 the LA Police department has been using its PredPol system to predict areas (500 feet square) where a crime is most likely to happen in the next few hours. http://www.predpol.com/. Accessed 6/04/2013.

2 Using predictive models

1. This is often for political reasons, because someone has a stake in certain data being required. Alternatively, people may feel reassured if the same type of data that was used in the previous decision-making system also features within the model.

2. Also known as leaf nodes or terminal nodes.

3. Often the score is taken to be the average value of cases that fall into each end node. For example, if 1,000 cases fall into node one, and five of these bought wine, then the response score would be 0.005 (5/1,000). This means that for classification problems, the score can be taken to be a direct estimate of the probability of an event. Likewise for regression: by taking an average, the score is then a direct estimate of the quantity being predicted.

4. For those of you who know how to build a predictive model, I would of course have used an independent holdout sample to produce the score distribution (or applied a more robust validation method such as leave-one out, K-Fold or bootstrapping), but for the sake of simplicity I have not discussed this here. Validation and ho dout samples are discussed in Chapter 8.

5. This is calculated as the average profit per case (Column F) multiplied by the expected number of responders (Column I).

6. Don't forget that this is only a prediction. It is very unlikely that the campaign will generate exactly 23,407 sales. 23,407 is our best guess. In practice, the actual number will probably be a few more or a few less than this because all predictions of this type have a degree of error. In the real world one would use some standard statistics to calculate confidence limits, showing what the expected range of sales would be.

7. Gross profit is $1,875,525 ($75 × 25,007). Total marketing spend was $1,198,898 ($200,000 + $998,898). Therefore, net profit is $1,875,525 − $1,198,898 = $676,627.

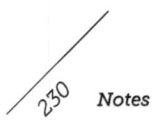

8. The actual model generates more scores than this, but for the sake of simplicity they have been grouped into three coarse categories.
9. This can be a bit confusing. The model built by the marketing team at Booles predicts the margin for those that respond. It does not predict the margin prior them responding. Therefore you can't use this model on its own to select who to target. You must also have an estimate of the likelihood of response as well in order to make accurate estimates of profitability.
10. Rapaport, Lisa (2012). *Amazon Tries Selling Wine Online For Third Time*. Forbes. http://www.forbes.com/sites/ups/2012/11/16/amazon-will-try-selling-wine-online-for-third-time/. Accessed 17/01/2014.
11. European Data Protection Directive 95/46/EC. At the time of writing this directive is under review and is likely to be superseded by revised legislation.
12. This is a fraud prevention measure to stop repeated applications from fraudsters who learn from each application and then modify their next attempt to obtain funds.

3   Analytics, organization and culture

1. International BASEL II/III regulations require each bank's assessment of its credit risks to be based on the use of quantitative analysis (i.e. predictive analytics) if it is to be allowed to use the preferred "Internal Rating Based (IRB)" approach to calculating its capital requirements. Similar requirements are contained in Solvency II regulations for the insurance industry.
2. Predictive Analytics World, London 2012.
3. In theory this is true, but even in academia you often find "positive result bias" – a paper is more likely to get published in a good journal if a positive result is found rather than a negative one.
4. This quote and a number of similar ones are attributed to a number of sources, but one of the earliest was by Publilius Syrus, a Roman writer and poet from the 1st Century BCE.
5. This is a very common practice. With credit card fraud, for example, when an individual has their card stolen or used fraudulently, the bank will simply close down the card account, refund the customer's loss and issue them with a new card. The only time a card issuer will involve the police is if they think the fraud is part of a larger fraud ring, involving many different accounts, and therefore, putting large sums (and/or their reputations) at risk.
6. In the countries in which the organization operated there is a requirement to give people an explanation as to why they are under investigation in order not to violate human rights legislation.
7. Or a two-model solution should be adopted, as advocated in Chapter 2. One model would predict the likelihood of fraud, the other the value of fraud when fraud occurs. The two models would then be combined to generate a single estimate of fraud loss.

8. In the UK, mortgage providers are required to take all reasonable efforts to avoid foreclosure. Before a UK court grants a repossession order, the lender must demonstrate that they have considered options for rescheduling the debt and/or have taken in to account any repayment options proposed by the borrower.

9. In most countries there are several steps that a mortgage company must take before foreclosure occurs, for example issuing a final statement and giving a number of days or weeks to settle the arrears before the case goes to court.

10. In the UK, about one third of repossessions are voluntary. Brownie, S. (2010). Research into Repossessed Households: Characteristics, Circumstances and Experiences. *BSA Seminar "Arrears and Repossessions."* 23/03/2010. http://www. docstoc.com/docs/82369163/brownie. Accessed 18/01/2014.

11. In most countries, when foreclosure occurs, if the sale of the property realizes more than the borrower's debt, then the borrower receives the balance of funds once their debt, plus legal costs, has been repaid. However, if the sale of the property does not cover the outstanding debt, the mortgage provider can continue to pursue the borrower for the outstanding amount.

12. GINI statistic (Sumer's D) of 0.84, KS-statistic of 0.62.

13. In 2012, the average foreclosure rate in the UK was about 0.4% (one in every 250) mortgages per year. The company in question had a somewhat better than average mortgage portfolio, with an annual foreclosure rate of around 0.2% (one in 500).

14. The actual review would take about an hour. The other hour and a half was required to prepare for the interview, review the paper work and undertaking any follow-up actions after the interview.

15. The model is built in stages. The preferred data items are allowed to enter the model before other, less desirable data items. This works because there is often some degree of correlation between variables, meaning that the variables that are allowed to enter the model first will feature more strongly (have more points or feature higher up the decision tree) than those that are considered later.

16. In some situations it is possible to allow overriding, but it has to be done in a controlled, well-managed way.

17. Much of this is based on "SWOT" analysis, where one considers the Strengths, Weaknesses, Opportunities and Threats associated with a project or business process. The development of SWOT analysis is attributed to Albert Humphrey, who developed the approach in the 1960s and 1970s.

18. The way the model was used would, of course, affect the volume of new customers, with some potential knock-on effects, but would not affect how customers where treated in any other areas.

19. Traditionally, there has always been friction between marketing and credit departments in banks. This is because the people most likely to respond to marketing activity for loans, credit cards, mortgages and so on tend to be the ones that are most risky. So the ones that the marketing department want to target (because the response rates are so high) are the people the credit department want to decline (because they are high risk).

20. The default risk from customers that don't repay their loans is a key determinant of the amount of capital that a bank has to hold to cover losses. Upgrading a bank's credit scoring models or changing the cut-offs used to make lending decisions will therefore have a big impact on the amount of capital that the bank must hold.

4   The value of data

1. A terabyte is 1,000 gigabytes. A gigabyte of storage can hold around 1 billion characters of text. This is enough for more than 1,000 books, several hundred pictures or about an hour of film.
2. For many years, the exemplar of data mining was how a supermarket had used advanced analytics to find a "hidden" relationship between beer sales and nappy (diaper) sales. As a result, the supermarket put these two items next to each other on their shopping aisles. The result was a big increase in beer sales. The reasoning was that men with young children stayed in at the weekend and would therefore also buy beer when sent out to buy nappies. However, this most famous of data mining stories is probably more myth than reality and there is little evidence to support it.
3. Kellison, B., Brockett, P., Shin, S. and Li, S. (2003). A Statistical Analysis of the Relationship Between Credit History and Insurance Losses, Bureau of Business Research. *The University of Texas at Austin.*
4. Two notable exceptions are crime, where a common objective is to identify the controlling entity at the center of criminal activity, and in marketing, where individuals at the center of a network are targeted in the belief that they will influence those around them to a greater extent than those at the edge of the network.
5. In the UK this information is held by the DVLA – a government department that maintains a register of all vehicles and their registered owners.
6. Hill, J. (2007). Structuring Unstructured Data. *Forbes.* http://www.forbes. com/2007/04/04/teradata-solution-software-biz-logistics-cx_rm_0405data. html. Accessed 17/01/2014.
7. This is true for most predictive models, where the objective is to predict future behavior. If one is predicting current or past behavior (such as someone's disposable income or who committed a crime) then historical data may not be required, or required to a lesser degree.
8. You may have a view about whether or not you think this is significant. However, if a formal statistical test was applied, the test would not be able provide a reasonable level of proof that a relationship existed.
9. In cases like this, where very little data is available, it's not possible to use standard mathematics or statistical techniques, but what you can do is create models based on expert opinion. These won't be as good as a statistical model built on a much larger sample, but these types of models can often provide a reasonable level of predictive accuracy.

10. This is a popular approach referred to as under-sampling. However, better results would be obtained from over-sampling. Over-sampling is where you take a larger sample of the majority class (say 50,000 in this example) and then balancing the population (i.e. weighting up the foreclosure cases).

11. This applies to most, but not all types of model construction technique including linear and logistic regression, CART and CHAID, which are the most common model construction methods. However, for clustering methods, for example, the relationship tends to be proportional to the square of the sample size.

12. This is for procedures like stepwise linear/logistic regression. If one undertakes singular (univariate) analysis of predictor variables one at a time, as a precursor to using a stepwise procedure, then this tends to show a more linear trend with the number of predictor variables.

13. This translates into a data set that is around 4–5 GB in size, prior to pre-processing it (which may result in a significantly larger file that is submitted to the modelling procedure used to generate the model). You can also do some trade-offs between number of variables and number of records. For example, if you only have twenty or thirty variables to consider, then a standard PC will have no problem constructing models using a few million records.

14. As at 2014, this would be something like a fourth-generation Intel i7 CPU based PC with 64 GB of RAM.

15. Crone, S. and Finlay, S. (2012). Instance Sampling in Credit Scoring. An Empirical Study of Sample Size and Balancing. *International Journal of Forecasting* 28(1): 224–238.

16. This is for classification problems. When I talk about 10,000 cases, this is 10,000 cases of each behavior. In this case, 10,000 examples of loans that went into default and 10,000 loans that were repaid.

17. In conjunction with tight entry/exit criteria for your model, i.e. data only features in your model if there is a very high degree of certainty (say 99.9% or more) that including it will improve the model.

5   Ethics and legislation

1. Aristotle, Brown, L. and Ross, D. (2009). *The Nicomachean Ethics (Oxford World's Classics)*, Oxford University Press, Oxford.

2. Chryssides, G. and Kaler, J. (1993). *An Introduction to Business Ethics, 1st edn*, Chapman and Hall. pp. 91–2.

3. Singer, P., Ed. (1994). *Ethics, 1st edn*, Oxford University Press. p. 243.

4. Immanuel Kant (1724–1804).

5. Boatright, J. (1999). *Ethics in Finance, 1st edn*, Blackwell. pp. 56–7.

6. Boatright, J. (1999). *Ethics in Finance, 1st edn*, Blackwell. p. 57.

7. It's interesting that almost no one disagrees with this principle when talking about a cake, but scale this up to society at large, apply it to income and assets

distribution, and the idea of us all getting the same is not something that many people agree with.

8. Olen, J. and Barry, V. (1989). *Applying Ethics, 3rd edn*, Wadsworth Publishing Company. p. 8.

9. Crisp, R., Ed. (2000). *Nicomachean Ethics, 1st edn*, Cambridge University Press. pp. 15 and 23.

10. Chryssides, G. and Kaler, J. (1993). *An Introduction to Business Ethics, 1st edn*, Chapman and Hall. pp. 101–2.

11. Olen, J. and Barry, V. (1989). *Applying Ethics, 3rd edn*, Wadsworth Publishing Company. p. 15.

12. Finlay, S. (2009). *Consumer Credit Fundamentals*. Palgrave Macmillan.

13. BBC News (2013). Corruption Getting Worse, Says Poll. BBC. http://www.bbc.co.uk/news/business-23231318. Accessed 18/01/2014.

14. Finlay, P. (2000). *Strategic Management. An Introduction to Business and Corporate Strategy*, 1st edn, Pearson Education Limited. p. 75.

15. Chryssides, G. and Kaler, J. (1993). *An Introduction to Business Ethics*, 1st edn, Chapman and Hall. pp. 249–54.

16. This example has been reproduced from one of my earlier books. Finlay, S. (2009). *Consumer Credit Fundamentals*, Palgrave Macmillan.

17. BBC News. 2013. Street View: Google given 35 Days to Delete Wi-Fi Data. BBC News. http://www.bbc.co.uk/news/technology-23002166. Accessed 11/7/2013.

18. In April 2011 the Sony PlayStation network was hacked, resulting in Sony closing down the network the following day. It was reported that personal information from over 70 million customer accounts had been stolen. It took almost a month for Sony to get the network back online following the attack.

19. In 2007 the UK government lost two DVDs containing personal details of all UK child benefit claimants – more than 20 million people. Child benefit is a universal benefit paid to almost every family in the UK with children.

20. Guardian, The (2013). Gang Stole $45m from Cash Machines Across Globe in Hours, say Prosecutors. *The Guardian*. http://www.guardian.co.uk/world/2013/may/10/us-crime-debit-cards. Accessed 18/01/2014.

21. ICO (2012). http://www.ico.org.uk/news/latest_news/2012/council-fined-250000-after-employee-records-found-in-supermarket-car-park-recycle-bin-11092012. Accessed 18/01/2014.

22. Department for Business, Innovation and Skills. (2013). *Security Breaches Survey*. Technical Report. Department for Business, Innovation and Skills.

23. How have I made this calculation? I have assumed that there are 100 million active credit cards in the UK. For each card I take the following information: Account number (8 bytes) expiry date (4 bytes), 4 digit Pin number (2 bytes), Security code on the back of the card (2 bytes), current balance (4 bytes), Date of birth (4 bytes), name (32 bytes) and address (100 bytes). Total storage requirement 16GB. This calculation is simplistic but reasonably accurate, and

in practice this data could be compressed to about a quarter of this amount; i.e. 4GB. With a modern USB interface this would only take a few minutes to transfer to a portable media device.

24. Lohr, S. (2010). The New York Times. http://www.nytimes.com/2010/03/13/technology/13netflix.html?_r=0. Accessed 23/7/2013.

25. Narayanan, A. and Shmatikov, V. (2008). *Robust De-anonymization of Large Datasets (How to Break Anonymity of the Netflix Prize Dataset)*. The University of Texas at Austin. http://arxiv.org/PS_cache/cs/pdf/0610/0610105v2.pdf. Accessed 23/7/2013.

26. Information Commissioner's Office (2012). *Anonymisation: Managing Data Protection Risk Code of Practice*. Information Commissioner's Office. http://ico.org.uk/for_organisations/data_protection/topic_guides/~/media/documents/library/Data_Protection/Practical_application/anonymisation-codev2.pdf. Accessed 18/01/2014.

27. The following is just one (highly cited), academic paper that provides strong evidence of positive outcome bias that occurs when medical research is sponsored by pharmaceutical companies which have a vested interest in the outcomes: Lexchin, J., Bero, L. J., Djulbegovic, B., Clark, O. (2003). Pharmaceutical Industry Sponsorship and Research Outcome and Quality: Systematic Review. *British Medical Journal 326*. For an interesting discussion about some of the reasons why such bias exists and the unethical practices that the pharmaceutical industry has been accused of, see the book *Bad Pharma*, by Ben Goldacre (2013).

28. A typical example of this relates to the UK, where payment protection insurance was proactively sold to pensioners and people on temporary contracts or who would otherwise not be able to make a claim on the policy. The UK banking industry has subsequently been forced to pay out billions of pounds in compensation payments to those who were mis-sold these products.

29. For one such example see: BBC News (2013). *SSE Fined Record £10.5m by Ofgem*. BBC News. http://www.bbc.co.uk/news/business-22011717. Accessed 18/01/2014.

30. You may be cynical about doctors making a packet, but regardless of their fees, all good doctors put the well-being of their patients at the top of their priority list.

31. I say mostly, because with almost every type of predictive model there are benefits to getting a subject expert to look at the model to spot any weaknesses and offer suggestions for improvement.

32. For example, a general bias correction method is Heckman's Bivariate Approach. Heckman, J. (1976). The Common Structure of Statistical Models of Truncation, Sample Selection and Limited Dependent Variables and a Simple Estimator for Such Models. *Annals of Economic and Social Measures* 5(4): 475–92. In credit granting (credit scoring) a specific process called "Reject Inference" is applied to correct for selection bias.

33. For an example of a highly cited study of name/race bias in recruitment, see Bertrand, M. and Mullainathan, S. (2004). Are Emily and Greg More Employable

than Lakisha and Jamal? A Field Experiment on Labor Market Discrimination. *American Economic Review* 94(4): 991–1013.

6   Types of predictive models

1. It's important to appreciate that predictive models are used to predict unknown events, and it's not a given that all unknown events are in the future. For example, when people use predictive analytics to identify credit card transaction fraud, the fraud is occurring now, in the present – it's not a future event. Likewise, using predictive analytics to come up with lists of suspects in a criminal case is looking to predict something after the event (the crime) that has already occurred.
2. Lovie, A. and Lovie, P. (1986). The Flat Maximum Effect and Linear Scoring Models for Prediction. *Journal of Forecasting* 5(3): 159–68.
3. Also called a parameter coefficient or parameter estimate.
4. In practice, there are a number of ways of determining how many bins there should be, and what range of values each bin should contain. Often binning will initially be performed using an automated algorithm. The resulting bins are then reviewed to ensure that the ranges make sense from a business perspective.
5. In theory, you only need four indicator variables for a problem like this because you can infer from the other indicators what the fifth one would be, i.e. an individual has only one income and therefore only one of the five indicator variables can take a value of one. All the others must be zero.
6. Another approach is to apply a cap. For example, for any income greater than $500,000 treat it as $500,000 for model building purposes.
7. Logistic regression is based on the principle of maximum likelihood. The likelihood is calculated as:

$$\text{likelihood} = P_1{}^* P_2{}^* \ldots {}^* P_G {}^* (1 - P_{G+1})^* (1 - P_{G+2}) {}^* \ldots {}^* (1 - P_{G+B})$$

where:
  G is the number of events in the development sample.
  B is the number of non-events in the development sample.
  $P_1, \ldots, P_G$ are the estimated probabilities of events, generated by the model, for events in the sample.
  $P_{G+1}, \ldots, P_{G+B}$ are the estimated probabilities of the event for each non-event in the sample.
  Therefore, $1 - P_{G+1}$ is the probability of non-event for the first non-event, $1 - P_{G+2}$ the probability of non-event for the second non-event and so on. The parameters of the model are chosen so that the likelihood is maximized.

8. That's the theory. However, one criticism of stepwise methods is that the assumptions upon which the tests of significance are based rarely hold true in practice. Therefore it's questionable as to whether all the variables selected are truly significant. Having said this, stepwise procedures work remarkably well in practice.

9. This is not absolutely guaranteed. It is still possible for significant correlation to exist, but correlation issues are greatly reduced.

10. This is generally, but not universally, true. Sometimes the contribution made by variables that entered at an early stage of the stepwise procedure diminishes as other variables enter the model due to correlations between the variables.

11. In theory, linear regression (least squares) is not appropriate for constructing classification models for a variety of reasons. In practice, however, the models generated using linear regression are pretty good, and if one is only interesting in ranking populations (rather than generating strict probability estimates), the performance of linear regression is comparable to logistic regression and other, more theoretically appropriate, classification methods.

12. It gets this name because it was developed after linear regression and shares many similar properties.

13. Quinlan, J. R. (1992). *C4.5: Programs for Machine Learning*, 1st edn. San Mateo, CA, Morgan-Kaufman.

14. Breiman, L., Friedman, J., Stone, C. J. and Olshen, R. A. (1984). *Classification and Regression Trees*, London, Chapman and Hall.

15. Kass, G. V. (1980). An Exploratory Technique for Investigating Large Quantities of Categorical Data. *Applied Statistics* 29(2): 119–127.

16. This is true, but often better (more predictive) decision trees result if some form of data transformation has been applied to standardized the data. Creating many indicator variables and then using these to generate a decision tree model is not recommended because it limits the splitting criteria that can be applied resulting in "sparse trees" that are not very predictive.

17. Rumelhart, D. E., Hinton, G. E. and Williams, R. J. (1986). Learning Representations by Back-Propagating Errors. *Nature* 323(6088): 533–6.

18. Common functions used to transform the score include the logistic (sigmoid) function and hyperbolic tangent function.

19. For regression problems it is common to apply the "Identity transformation." This just means that the score is not transformed at all, i.e. the raw score calculated by the output layer neuron is the final score, without any transformation being applied via the activation function.

20. With back-propagation the data scientist needs to select two control parameters. These are the learning rate and the momentum, with the best values for a given problem often found only after much trial and error. Methods such as Quasi-Newton and Levenberg–Marquardt optimization don't require these parameters to be set and are many times faster than back-propagation.

21. With K-means clustering, K cluster centers are defined. Each observation is then clustered with the cluster center closest to it. With hierarchical clustering, each observation is clustered with its closest associate, i.e. the observation to which it is most similar. So at this point you have lots of small clusters. You then repeat the process a number of times, each time joining together the clusters that are most similar, so that by the end of the process you just have a few large clusters.

22. Although generating linear type solutions via weighting variables is the most common way in which the Delphi method is applied, in theory there is no reason why it can't be applied to other types of model – in particular decision trees, with the group deciding which variables to use to segment the population.

23. In credit scoring, the most widely cited research study of the predictive accuracy of different methods, and which finds very little difference between them, is: Baesens, B., Gestel, T. V., Viaene, S., Stepanova, M., Suykens, J. and Vanthienen, J. (2003a). Benchmarking State-of-the-Art Classification Algorithms for Credit Scoring. *Journal of the Operational Research Society* 54(5): 627–35. Likewise, for similar results for direct marketing see: Linder, R., Geier, J. and Kölliker, M. (2004). Artificial Neural Networks, Classification Trees and Regression: Which Method for Which Customer Base? *Journal of Database Marketing & Customer Strategy Management* 11(4): 344–56. Another such study for the insurance industry, which again comes to similar conclusions is: Viaene, S., Derrig, R. A., Baesens, B. and Dedene, G. (2002). A Comparison of State-of-the-Art Classification Techniques for Expert Automobile Insurance Claim Fraud Detection. *Journal of Risk and Insurance* 69(3): 373–421. For a more general study that looks across a variety of datasets from a range of different application areas see: King, R. D., Henery, R., Feng, C. and Sutherland, A. (1995). A Comparative Study of Classification Algorithms: Statistical, Machine Learning and Neural Networks. *Machine Intelligence* 13. K. Furukwa, D. Michie and S. Muggleton. Oxford: Oxford University Press.

24. This is because the things that predict many types of consumer behavior can be expressed relatively simply and/or are linear in nature. Using a complex algorithm is like using an expensive piano to play "Three Blind Mice," and can introduce what is termed "noise," i.e. spurious relationships are captured by the model resulting in worse performance than a simpler representation of the problem.

25. Interaction is where the relationship between a predictor variable and behavior changes, depending on the values of other predictor variables. For example, higher income may be predictive of being more likely to own your own home rather than renting it. However, the inverse relationship is seen when we are talking about high-value properties in city centers (such as penthouses in New York or London).

26. A model developer would also consider the use of interaction variables – but that's getting into some of the more technical aspects of model building.

27. Hand, D. J. (2005). Good Practice in Retail Credit Scorecard Assessment. *Journal of the Operational Research Society* 56(9): 1109–17.

28. This includes linear models where a binning strategy has been applied, even if the raw data displays considerable non-linearity.

29. This is because of the inefficient nature of common splitting algorithms used to derive decision trees. Each time the population is split, the resulting segments have a smaller number of cases than their parent segment. Therefore, there are

only a finite number of splits that can be applied before one runs out of cases. As a rule: if the minimum node size to avoid over-fitting is N and there are M variables that should feature n the model, then at least $N \times 2M$ observations are required to fully represent the problem using a decision tree. For example, if we have 20 variables that are predictive, and the minimum node size is 100 observations (which empirically is about the number you need for most predictive analytics problems) you would need at least $100 \times 220 = 104,857,600$ observations to fully capture the behavior. This compares to around 2,000 observations that would be required to capture the same behavior with a linear model or neural network.

30. It is generally taken to be good practice to have equal numbers of cases of events and non-events, regardless of the type of model being constructed. However, decision tree algorithms are far more sensitive to imbalanced data sets than all of the other methods discussed in this chapter by a considerable margin. See the following paper for some empirical analysis of the effect of class imbalance: Crone, S. and Finlay, S. (2012). Instance Sampling in Credit Scoring. An Empirical Study of Sample Size and Balancing. *International Journal of Forecasting* 28(1): 224–238.

31. The two popular methods for creating a balanced sample are over-sampling and under-sampling. With over sampling cases from the minority class are replicated a number of times so that the two classes contain (approximately) equal numbers. For example, if you are building a model to predict mortgage fraud using a sample that contains 1,000 frauds and 250,000 non-frauds, each of the frauds would be replicated 250 times. With under-sampling cases from the majority class are "thrown away," i.e. 1,000 cases of non-fraud are selected at random to create and equal (balanced) sample. As a rule, models constructed using over-sampling outperform models constructed using under-sampling.

32. If you consider the spend models discussed in this chapter, for the two linear models there are hundreds of different scores that can result, based on different combinations of the data. However, for the decision tree, there are only twelve possible scores.

33. Regulation B of the Equal Credit Opportunity Act 1974 (Often referred to as the ECOA).

34. With some clustering software you do get quite a lot of information about how the cluster was formulated. For example, with hierarchical clustering most software will show you how the initial clusters were formed, and then how these were merged together to produce the next set of clusters and so on.

35. For example, most of the world's credit scoring models are only redeveloped every one to three years.

36. Hand, D. J. (2006). Classifier Technology and the Illusion of Progress. *Statistical Science* 21(1): 1–15.

37. I would also recommend the use "tight" entry criteria for the stepwise procedure to reduce over-fitting The entry criteria relates to the certainty one has that the variable truly contributes to the performance of the model; i.e. the observed relationship is not due to random effects. For explanatory models, it is usual to

construct models using 90%, 95% or 99% significance. For predictive modeling it's usually better to be more certain than this. My recommendation is to explore the use of significance levels of between 99.00% and 99.99% My experience is that if you use a lower level of significance then more variables feature in the model, but these additional variables don't add much, if anything, to the predictive ability of the model and often results in a degree of over-fitting.

38. Breiman, L. (2001). Random Forests. *Machine Learning* 45: 5–32.
39. The standard process for building a random forest is as follows:

    1. Create a new development sample, by randomly select cases from your original development sample (via random sampling with replacement; i.e. Bootstrapping).
    2. Randomly select a few (four or five) of the available predictor variables.
    3. Build a simple decision tree using the development sample and predictor variables that were selected in steps one and two.
    4. Repeat the process several times.
    5. Combine the outputs (the scores) from all the trees to generate a final prediction.

    Due to the randomness of the process, each model is constructed using a slightly different development sample and a different set of predictor variables. Most applications of random forests develop somewhere between 25 and 200 trees/models. However, some applications have 10,000 separate trees or more.

40. Schapire, R. (1990). The Strength of Weak Learnability. *Journal of Machine Learning* 5: 197–227.
41. Friedman, J. H. (1999). *Greedy Function Approximation: A Gradient Boosting Machine,* Technical report, Department of Statistics, Stanford University.
42. Siegel, Eric. (2013). *Predictive Analytics: The Power to Predict Who Will Click, Buy, Lie, or Die,* Wiley. p. 148.
43. For some examples (relating to credit risk) where ensembles did not add value, or the value was only very marginal see:

    1. West, D., Dellana, S., and Qian, J. (2005). Neural Network Ensemble Strategies for Financial Decision Applications. *Computers & Operations Research* 32: 2543–559.
    2. Zhu, H., Beling, P. A. and Overstreet, G. (2001). A Study in the Combination of Two Consumer Credit Scores. *Journal of the Operational Research Society* 52(9): 974–80.
    3. Banasik, J., Crook, J. N., and Thomas, L. C. (1996). Does Scoring a Sub-Population Make a Difference? *International Review of Retail Distribution and Consumer Research* 6: 180–95.

44. A 10% improvement was defined in terms of the reduction in the error of predictions. The error measure used by Netflix was Root Mean Square Error (RMSE). To win the prize the competitor's model would need to reduce the RMSE from 0.9525 to below 0.8572.

45. Amatriain, X. and Basilico, J. (2012) Netflix Recommendations: Beyond the 5 stars (Part 1). *The Netflix Tech Blog.* http://techblog.netflix.com/2012/04/netflix-recommendations-beyond-5-stars.html. Accessed 23/6/2013.
46. The concept of least squares, which is the foundation of linear regression, dates to the first decade of the 19th century. The idea of extending the least squares principle to apply to many predictor variables arose towards the end of the 19th century/start of the 20th century based on work of the Polymath Francis Galton and the Mathematician Karl Pearson.

7   The predictive analytics process

1. PRINCE 2 is a project management approach developed in conjunction with the UK government in the 1990s, specifically to manage large public sector IT projects. The methodology of PRINCE 2 has been widely adopted by many organizations undertaking large-scale projects.
2. I have never yet come across a problem where it would be appropriate to consider all of the data that an organization can access to build a model. There is always a huge amount that can and should be discarded. Often this is due to the age of the data (too old or too recent) or because that data does not relate to the behavior that you are trying to predict.
3. For clothing retailers, returned items are a big issue. It's not unusual for around 30–40% of clothing items sold remotely (via Internet or mail order) to be returned, representing a very significant overhead for the retailer.
4. It is possible to build multi-class (multi-nominal) models that predict more than two classes of behavior, but these tend be less popular with practitioners than binary classification models.
5. The terms "good" and "bad" are standard terms used across the credit industry to define customer repayment behavior.
6. In the credit scoring community these types of cases are termed "Indeterminates" and are excluded from the construction of credit scoring models.
7. In these examples, I've focused on the objectives for classification problems. However, the same issues arise when one tries to model continuous outcomes, i.e. when one builds regression models.
8. At one level this process removes errors during the transfer process. However, there could be differences between the way data items are calculated in the operational system and how they were calculated on the analytical server. Therefore there remains a requirement to test the model to ensure correct implementation before it is put live.
9. Statins are drugs that reduce the production of cholesterol by the liver. Cholesterol is a key factor in many types of heart disease.
10. Of course the doctor could have several models: one to predict long-term risk that is used to decide preventative treatments, and another short-term model to drive a different treatment plan in response to the impending heart attack.

11. It can be a bit more complex than this when a model is being redeveloped because new sources of data become available. In this case, as well as building the new model you will also need to amend your IT systems to enable the new data to be included in the scoring process.
12. With this sort of process it's always better to identify problems sooner rather than later: for example, it is often very difficult to recover incorrect benefit payments once they have been paid out. Therefore there are a lot of savings to be made by preventing incorrect payments before they occur.
13. It often takes this long because the consultant won't have any understanding of the data and therefore, needs to learn about it. There will also be several meetings with the client to discuss requirements and deliver results.

8 How to build a predictive model?

1. 121,929 Ford Fiestas where sold in the UK in 2013. The Society of Motor Manufacturers and Traders (2014). http://www.smmt.co.uk/2014/01/2013-new-car-market-records-best-performance-five-years/. Accessed 20/01/2014.
2. This type of thing usually arises due to a system change, or when two separate databases are merged – for example, when a company takeover occurs.
3. Calculated as a person's weight in kilograms divided by their height in meters squared. In the UK, a Body Mass Index of 18–25 is considered normal, 26–30 overweight, and over 30 obese.
4. Missing data is a huge problem for all sorts of analysis. A key factor is if the data is missing at random, or missing due to some specific reason, in which case the reason why the data is missing is in itself an important piece of information. In this particular example, it may be the case that the data is perfectly valid for the 1% of cases where residential status is available and therefore it can be used for model construction. However, understanding why it's only available for 1% of cases is important.
5. In England, a County Court Judgment (CCJ) is the standard legal procedure that creditors use to recover an unsecured debt. A CCJ is far less serious than a bankruptcy petition as it does not result in forfeiture of the debtor's assets or loss of their home. If the ruling is upheld, then the court will order the debtor to repay the debt and a repayment schedule will be agreed. If the debtor fails to meet the repayment schedule, then bailiffs can be appointed to recover goods in lieu of the debt.
6. The most popular measures for classification problems are: Information Value, Efficiency (GINI) and the chi-squared statistic.
7. For decision trees the use of indicator variables is not recommended as it results in sparse trees whose performance is poor.

8. For Weight of Evidence (WoE), a number of ranges (bins) are defined, in just the same was as you would if you were creating a set of indicator (dummy variables). The WoE for each range is defined as:

$$WoE = Ln\left(\frac{g_i \, / \, G}{b_i \, / \, B}\right)$$

where:

g_i is the number of events within the range.
b_i is the number of non-events within the range.
G is the total number of events in the whole population.
B is the total number of non-events in the whole population.

The WoE is then substituted for the actual value of the variable for each case. WoE acts to standardize and linearize the data, and has the advantage that only a single transformed variable is created (Instead of several indicator variables). The disadvantage of WoE transformed variables is that they are less flexible than indicator variables and usually result in a marginally worse (less predictive) models.

9. I don't believe that this tool lets you see the model that has been constructed: it's very "black box" in nature. You present it with the data, it builds a model and then lets you score new cases through the model, but without letting you see the actual model that has been constructed.

10. In most countries men still earn more than women, even if they are doing similar jobs. Legislation to prevent discrimination on the basis of gender only has a limited effect on salary differentials.

11. This was a management failing far more than it was a mistake by the poor analyst. A more experienced team member should have gone through the presentation with the analyst beforehand and suggested a different approach that would have made it more suitable for the target audience.

12. Use of predictive models is strongly encouraged for Advanced Internal Rating Based (A-IRB) approaches to credit risk assessment.

9 Text mining and social network analysis

1. Many stop word lists also include pronouns such as "I," "he," "her," "it," "them" and so forth. However, for advanced text analytics it is a mistake to remove pronouns as they assist in understanding the holistic meaning and subject matter of the text.

2. Just using these rules will not get it right every time. In particular, many nouns will end up being incorrectly truncated. For example, Trees, Grass and Wedding, would end up as Tre, Gras and Wedd, after the application of these stemming rules.

3. The correlation approach most commonly used in text analysis is cosine similarity. For each document each word is represented as an element in a vector, where each element (dimension) of the vector captures the number of times the word appears. Cosine similarity then gives a useful measure of how similar (how correlated) the two documents are.
4. This is a fabricated example, only very loosely based on my real family.
5. For the record, this is not a practice I am in favor of.

10   Hardware, software and all that Jazz

1. A power dialler is a system that is loaded with people's contact details and then automatically phones them. When someone picks up the phone the dialler transfers the call to a human operator.
2. The first in-database analytical system appeared in the 1990s, but they have only come to prominence since the late 2000s.
3. So that's one million cells containing product ID in the Sales Table, plus 6 × 20,000 cells in the Product Table.
4. A fuzzy match is where two data fields are considered to be the same, even though they display slight differences. Imagine we have name, date of birth and social security number data in two records. Record 1 contains "Steven Finlay 04-08-1974 111-222-333" and Record 2 contains "Stephen Finlay 04-08-1974 111-222-393." You and I would say that these two people are probably the same. Somebody has just mis-spelt Steven/Stephen in one of the fields *and* got one digit of the social security number wrong. An exact matching system would not consider these two to be the same unless they were identical in every respect.
5. There are many different implementations of SQL. There are core standards that vendors of each version adhere too, but many implementations of SQL, such as MySQL and PostgreSQL, offer additional functionality.
6. Wikipedia. http://en.wikipedia.org/wiki/Apache_Hadoop#cite_note-8. Accessed 20/01/2014.
7. A cheap server can cost just a few hundred dollars. A top of the range model can cost several thousand dollars, but this is for the hardware only. You'll also need the software to sit over the top of it, and then pay someone to connect it up so that you can use it. As a rule of thumb expect to pay twice the cost of the hardware to get a server system up and running.
8. RAID stands for Redundant Array of Independent Disks. RAID is a standard way of protecting data by storing it across several different disk drives.
9. For more information about HPCC visit the website: http://hpccsystems.com/.
10. For more information about Storm visit the website: http://storm-project.net/.
11. https://cwiki.apache.org/confluence/display/Hive/Tutorial.
12. Appuswamy, R., Narayanan, D., Gkantsidis, C., Hodson, O. and Rowstron, A. (2013). Nobody Ever Got Fired for Buying a Cluster. *Microsoft Research, Cambridge, UK. Technical Report.*

13. I've assumed four bytes to store a product code for each item purchased, which equates to about 100GB. I've then assumed 40GB is required to store the other 1,000 items of data for each customer. In reality, there would also be some additional data about visit times, item prices and so on, but this would add only marginally to the overall size of the database. Consequently, I've allocated a nominal 10GB to cover such things to get to my 150GB total.

14. What some people argue is that if I take a sample that's 10 or 100 times larger than this, then additional variables will feature in the model. This is because with a few more cases it's possible to apply statistical tests that prove beyond reasonable doubt that there is an association between that predictor variable and behavior. However, statistical significance is not what's important when it comes to predictive analytics. The magnitude of the effect is what matters. If a predictor variable is statistically significant, but it only affects a fraction of 1% of the population by a very small amount, then it's not going to add much to the bottom line.

15. This process of grouping bins together is sometimes referred to as "coarse classing."

16. There are various business reasons why you might want to do these two things. For example, you may want to force certain variables to feature in the model to encourage the business to accept the model. Likewise, it may be politically expedient to exclude certain variables from a model.

17. PMML standards are maintained by the data mining group that represents vendors offering solutions that support PMML. http://www.dmg.org/products.html. Accessed 20/01/2014.

Index